A Bus Will
by
Johnny Bock

Lunchbreak Press
PO Box 1971
Eagle River, Wisconsin 54521-1971
(lunchbreakpress.com also
Lunchbreak Press on facebook.com)

©Copyright 2017 by Johnny Bock
All rights reserved.
Printed in USA

Other books by the same driver

A Bus Will Take You There

The Reference Point: A Journey to the Origin of Belief

The 99th of August

The Sunstone Furnace:
A New Way to Heat with an Old Fuel

To those who have left the larval stage
and are ready to try their wings.

Also for "Crusty" and Dawn, who will always be
remembered, but will never be replaced.

Contents

Echoes from Days Gone By	1
New Growth All Around	7
A Big Dog	21
You Can't Stop Progress	54
Four Dogs and Two Wars Later	77
Do What Must Be Dung	86
Function	106
A Blank Slate for The Driver	118
End of Report	139
Check the Box	161
In Good Hands	170
Sacred Ground	183
Winter Will Do That	196
More Than Circuitry and Silicon	206
The Dual Nature of The World Outside	237

Somewhere Inside

"What the road don't give you the miles can't."
           —Bobber

"No, Dad. You don't understand. I hafta go right now!"   —Mary Teal

"I don't know how the law would see this in your state, but here in Illinois we're talking a ten thousand dollar fine and three years in jail"
         —Officer Stickler

"My chest hurt, and when I breathed I could feel crackling inside."   —Sid

"You will trade your Bus for a horse, Jimmy, and leave behind those stone rivers that keep taking you away."   —Douglas

"It's not there now but it keeps coming—there, there it is again. That."   —Beano

"I looked up and saw Steve's hand reaching down."   —Judelaine

"I have the gun. You can pay me in beer."
            —Ejo

"Here's where I take a lesson from the wind. Always be quiet unless you are only passing through."   —Jimmy Yellowbird

Echoes from Days Gone By

"For fulfillment, prophecy needs none of your help. What it needs is your opposition. Only then can it prove to be truly prophetic."

(From the first draft of the Book of Exodus, edited out under pressure from the travel industry.)

The louder the shouts for time to stop, the harder to hear the sounds from tomorrow. Most of the sounds from tomorrow only call for minor additions and easy corrections, but sometimes the message is Definite Radical Change. You can always tell when the sounds are warning of radical change when the ones who shout for time to stop only make the night last longer.

Who would think the 1950s would give birth to a time of radical change? The 1950s? To the people living then, the 1950s were seen as the Dawn of the Age of Eternal Suburbia. You got a job in the city, a house in the suburbs, and raised 2.3 kids—all with the understanding that those kids would grow up to get a better job in the city so they could buy a bigger house in the suburbs and raise 2.4 kids.

It certainly made sense at the time. Why keep going once you've arrived?

But sounds from tomorrow also arrived, and some were heard in the form of a driving beat. Kids called it rock music, and their parents called it a phase. Just like crop-circle haircuts and ridiculous clothing, the parents thought, this attachment to rock music was only a temporary turnout on the road to maturity. An annoying turnout, to be sure, and sometimes so annoying as to cause the parents to shout, "Turn that shit off!" But, they thought, once the kids hit their stride—once they hit the age where they see the value of Life in Suburbia—Mom and Dad generally believed that the kids would drop their phony rebellions and shed their silly songs and take that and all their other teenage trappings out Tuesday with the trash.

Good luck with that.

Because by the time the teens from the '50s were due to grow up—right about that time, real reasons to rebel were making themselves heard. If you listened to the news, one of the reasons was cooked-up American aggression halfway around the world, and if you listened to your conscience, another reason was American repression right here at home.

And right here at home it broke out as loud arguments between one generation "too young to understand how the country really works" and another generation "too old to know that in order to keep working, the country has to live up to its promises."

Civil rights and Vietnam.

You remember. Unless you supported the sequel to the Vietnam War, commonly known as the War For Oil And Dad, you remember very well. People were pissed. The ones "too young to know how the country really works" were not going to settle for a standard little self-centered rebellion featuring funny haircuts and ridiculous clothing. No, theirs was going to be bigger than that, and they were going to show it by breaking "rules" and making a lot of noise until they got some satisfactory answers for questions like "Where is the sense in sending sons overseas to die in a war whose stated purpose is to produce a 'body count'?" and "What kind of a democracy would threaten its citizens with death or jail when all they did was try to vote?"

No doubt about it. The young—regardless of their age—were plain up-to-here with what they saw as the overfed undernourishment of that segment of superficial society affectionately known as "their parents." Maybe their parents were members of the elder generation, the young were saying, but they certainly weren't a Generation of Elders.

This is where time took on a new identity. Suddenly it was the '60s. The placid '50s were gone and there was no going back. Loud and radical seismic change went from arguments at the dinner table to violence in the streets, and for most of the decade social attitudes and cultural foundations would be shaken with all the ferocity and intensity of a dog shaking a rat.

But somewhere in the middle of all the sirens screaming and bullhorns blaring—somewhere east of the riots and north of the gunfire—came another sound. It caught your attention not for its high volume or fanatical insistence, but for its gentle suggestion. The breezy way it sounded suggested that maybe it was time to lighten up. Maybe it was time to put light on the struggle instead of noise, and such a suggestion had an immediate appeal to those who had no urge to kill their way to peace or burn their way to brotherhood.

It was the sound of a little air-cooled engine.

To hear it run made you instantly curious. "What is that?" you would ask, and when you discovered the source you were both amused and amazed. "Is that a car?" you would say, smiling, and not ready to accept that what you were looking at was actually automotive. Instead, it looked more like a bug, and if you wanted to be more specific, that bug would be further described as a "beetle."

Officially though, the name of the little car was "Volkswagen." That's all. Just Volkswagen. Not Volkswagen "Barracuda," not Volkswagen "Fury," and not Volkswagen "El Dorado, Imperial, Park Avenue or Pork Avenue." It had none of the subtitles of the top brands of the day. Nothing intending to inject the driver with feelings of power, dominance, or rage, and no hint of luxury, opulence, or other forms of automotive muscularity or vehicular obesity. Just Volkswagen—which in German meant "People's Car."

Some had a problem with that. Not the term "People's car," but the word "German." For them, it was still too close to World War Two, and at this point in history anything coming out of Germany reflexively registered as a military invasion.

Almost as a counterclaim to that kind of thinking, Germany followed up on the Beetle with the Bus. A country with a well-deserved reputation for exporting mass murder and industrialized vandalism unknowingly created and exported what was to become the peace-sign bearing "hippy van." Unknowingly, because conceptually, the Volkswagen Microbus was built strictly for utility. Originally called the Transporter, the Bus was little more than a rolling box with three pedals, a gearshift, and a steering wheel. Also there was a seat. As bleak and utilitarian as the early models were, you'd think the design would have the driver sitting on a tool box bolted to the wheel well, but no, there was actually a seat.

Now, for a person not one bit impressed with vehicular muscularity, and for a person thoroughly disgusted with automotive obesity, a rolling box with not much more than three pedals and a seat seemed just about right. "Look at that," I softly said the first time I peered into the interior of a Volkswagen Microbus. Nobody was around to hear, but it didn't matter because "Look at that" came out as a quiet exclamation of discovery. It came out as a more elevated and less juvenile way of simply pumping a fist while letting out a burst of happy profanity because "Look at that" meant I could stop looking and start buying.

My sense of discovery was not a feeling widely shared. If one person nodded in casual agreement, ten others taking a peek into my newly bought Microbus—friends, even, even relatives—and close relatives—saw only three pedals, a gearshift, and an idiot. They were quick to shake their heads and were just as quick to comment that anyone who would pay good money for such an empty box of nothing "must be pulling on his pants with the brown in the front and the yellow in the back."

But whatever lacks in finish is loaded with possibility. After I put the title in my pocket, and before I put the key in the ignition, I took one more look into the interior of the Bus. For such a small vehicle, it had a lot of room. And almost all of the room was room for possibility. The future has a lot of room too, and has a lot of room for possibility. So if you took the future and put that also in the Bus, then there would be even more room yet, and all of that room together would amount to a continuing expansion of ever greater possibility.

The former sounds from tomorrow are now heard as echoes from days gone by. We live, and with memory we live again. So brew up a cup of coffee, and when the drink is gone, hold the mouth of the empty cup near your ear. Do you hear the sound of tires rolling over a distant gravel road? Is this a sound of meaning both here and there? If so, take the empty cup away, relax, and A Bus Will Bring You Back.

New Growth All Around

"There is what a person is, and there is what a person does, and the two of them together are what a person becomes."

(From the journal of a woman recently divorced from a man whose job it was to pump septic tanks.)

When I came back, three of them were eyeing up the Bus. Soon there would be more, many more, but so far only three, with each having those kinds of thoroughly muscular builds that show bulging biceps even with the arms extended.

Under a streetlight in the bad part of town, such a scene could be seen as a threat, but this was mid-day in front of a mini-mall. It was a clean little mini-mall located just off a busy street in the middle of a middle-class neighborhood, and there, lined up one next to the other was a grocery store and a video rental, a coffee shop and a fitness center. Also a pizza parlor and a hair saloon. In other words, safe, and a good place to pack in a few food items before motoring up into the tall timber to spend some quality and quantity time in close company with the lakes and the loons.

All the parking spots were taken in front of the grocery store, as were the spaces in the adjoining business, which was the hair saloon, so I kept on going until finding one in front of the fitness center.

Since it was a warm day in early fall, the doors of the fitness center were open and folded back. They were double doors, wide, and while walking past them you couldn't help but notice the sounds of feet plodding on treadmills, the clank of barbells, and the whirring of cables racing through pulleys.

What you also couldn't help but notice was the presence of a certain scent. Along with the sounds, wafting through the doorway was that unmistakable scent given off by concentrated muscle effort. It was an agreeable kind of a scent that you sensed would remain agreeable as long as the responsible bodies were washed beforehand, but let those bodies end their activity without another washing, and soon such a scent would cease to be a scent and would assume the role of an odor.

Not all the time, but too many times for the cultured and the sensitive, I put the snidemouth on the motives of others. You know how it goes. Unless you're brain deaf, blind in one eye and wearing a patch over the other, you know how it goes. As an example, say somebody does something for a good reason. Then a guy like me comes along, sniffs at it, and claims they're doing it for a reason other than good. You know how it goes.

They say that kind of commentary is common. They say everybody does it, although you'd never know by the vehemence of the denials. My own way of denial is to say I'm only doing it to be funny. All joke, Mon, b*eeg* joke. Like when I passed the open doors of the fitness center, instead of nodding in approval at their commitment to a better way of being, I thought of them as caged hamsters and wondered how many miles per head of lettuce they could get while running on the treadmill.

It got worse while passing the hair saloon. And you probably knew this was coming by seeing the word "saloon" put in a place that should be occupied by

8

"salon." Anyway, while passing the hair saloon, and while taking in all the pampering going on inside, instead of giving the hairdresser a heads-up for making pretty women beautiful, I thought of her as a saloon keeper dispensing vanity in place of whiskey, and by this act was keeping the customers away from what they should really be doing, which would be carrying the family laundry down to the river, barefoot.

> Disclaimer: that's not me talking. That final utterance in reference to the laundry and the river and the bare feet came from the oink of a little chauvinist piglet living in my head. He never pays the rent, so sometimes, just to expose him for what he is, I let him say things he will regret later.

Hoping to keep the little devil from saying something he would regret later even more, I made up my mind to walk straight to the Bus after leaving the food store. No sideways glances and no chances of the pesky little peckerneck slipping in a snide remark that might make people think he was somehow associated with sweet and saintly me.

Which turned out to be easy, because straight ahead, there they were, the three big musclemen, eyeing up the Bus.

"This yours?" one of them asked when it was obvious to them that the guy lugging the groceries was lugging them to the Bus.

"Well," I said, stopping among them and taking in the amount of movable meat I was up against if things went south, "If you like it I'll claim it, but if you don't, then I'm just taking it to the junkyard for a friend who's too embarrassed to take it there himself."

"No, I was just saying. A friend of mine has one just like this. Same color and everything. Is this a '66?"

"A '62, actually. But from something like 1853 to 1967 they all look like they came from the same mold."

"See, I told you so," a second muscleman said, standing next to the first. "I kn*e*w it was older than Nate's"—and at the word "kn*e*w" he gave the first muscleman a quick backhanded slap on the shoulder.

The playful quality of the slap made it clear that the two were friends, and that slapping seemed to be an accepted way of showing it. Me though, new to the scene, and carrying among other things, eggs, I didn't want to get caught in the middle of a friendly slapping escalating into a passionate punching, kicking, and shooting, so I quickly said, "Technically you're both right. The two front doors are off a '66 and so is most of the glass and half of the driver's side. Along with part of the roof."

When they just stood there looking at me I went on to explain how the Bus had been totaled twice, and how each time it needed a lot of medical attention and boocoo life support, spending months in intensive care, requiring a long list of transplants involving bone, skin, and more, and that most of the transplants came from a '66. "Look here, for instance," I said, opening the cargo doors. "You can still see some of the welds from the inside. That bead near the bottom on the other side goes way back behind the second window, and you can't see it now because of the upholstery, but then that same bead goes up and over the middle of the roof and continues on down to the front."

At that, the third muscleman, quiet till now, turned and hollered into the fitness center, "Hey Shrader, take a break! You gotta see this!"

At first I thought maybe Shrader was a welder and that the third muscleman meant to have him check out the beads, but no, when Shrader arrived, the third muscleman gestured to the inside of the Bus and said, "Can you believe it? Look at all the sh*it* in there."

There was a long moment of silence while Shrader leaned in and looked at "all the shit in there." Finally, straightening up and giggling, he turned and took two steps into the fitness center. "Ted!" I heard him holler. "Get out here! Quick!"

The urgency of his tone brought more than Ted outside, and brought them out in time to hear Shrader say to Ted, "The Swiss Army Bus."

"What?"

Shrader ushered Ted over to the Bus, and giggling again, motioned to the inside and said, "You heard of the Swiss Army Knife. This is the Swiss Army Bus."

With Ted and Shrader looking into the Bus, and with others looking in over their shoulders, I couldn't see any faces, but even at that it was safe to say that the faces were not faces of disappointment because I kept hearing comments like "Kick ass!" "I gotta get me one a these," and "Too bad Ryan's not here. He needs to see this."

Also somebody said, "I wonder if those chairs are legal," and somebody else answered with, "Yeah, and what about the stove. You can't tell me that isn't dangerous."

The rising speculation finally got to the point where there was a desire for some solid answers, so when someone asked, "Whose is this, anyway?" one of the original musclemen pointed at me and said, "His."

Since this wasn't the first time the Bus was an object of curiosity, most of what they wanted to know I already had answers for. So, when someone pointed at the stubby-legged little stainless steel box parked on the floor behind the driver's seat and said, "Is that a stove?" my ready answer was, "Actually it's the third one. I burned two of them out already, and this one here's got about six years on the clock."

There's something about a wood-burning stove in a licensed vehicle that runs counter to what your average person can believe. At the sight of the little stove, with its glass door, side-shelves, and one-piece chrome stack vented through the roof, people have a tendency to treat it as an illusion. They look, and they blink, and after it's still there when they blink again, then they're ready to accept it, but only as a decoration. They're ready to accept it as a quaint little combination of art and craft, but certainly not as a thing of working utility. That's why I was not one bit surprised when one of the onlookers flatly said, "No way does that thing work."

Since truth can only trump belief with proof, here's where I opened a trap door in the floor of the Bus and said, "Too bad it's too early for fishing on the ice. We could drive out on the lake, drill us a hole in the ice down through this hole in the floor here, and after we'd drop in a line and pull out some fish we wouldn't have to worry about going hungry because the stove would be hot and ready to fry up everything we caught."

"You mean you can cook right here on the stove?"

"In the stove too." I pointed to the grilling fork clipped nearby and went on with, "Stick that thing inside the stove with a steak on the tines, and before you know it, what you've got is something that goes pretty good with a bottle of beer."

After more of the same kind of talk, and after opening the door of the stove to expose ashes from the last fire, it was finally accepted that yes, it was a wood-burning stove, and yes, it cooked and fried and grilled, but more than that it had the basic advantage of being a battle-tested soldier in the war against frostbitten fingers and toes.

Harder to accept were the seats. "No offense," someone said while scoping out the front seats, "but my brother's a cop, and I bet if he saw those lawn chairs there, he'd have this thing towed."

"Them ain't lawn chairs," I said. "Them's webbed seats."

"Seriously though. Again no offense, but from a safety standpoint—I mean what would happen in a crash?"

"Well," I said, taking it from there. "I don't know how these chairs here would do in a crash, but the last ones of pretty much the same design let the driver and his rider walk away from a wreck that was meant to put them both six feet under."

Those words were followed by the recounting of the time some guy in a truck deliberately hit my son and me from behind, pushed us down the highway, and at a speed more than the posted limit made us lose control on a curve, where from there we flopped, flipped and rolled, and finally came to a soft landing in a snowy ditch. The whole thing turned the chairs into pretzels and the Bus into junk, but my son and I stepped out alive and whole. We were bumped and bruised, but the biggest hurt was not being able to run down the offending vehicle and do wreckies on the driver.

"You mean somebody actually hit you on purpose?"

"Not only did he hit us on purpose. He hit us in the back on purpose. And then he ran away."

This brought out a round of murmurs and more questions. "Did you get a look at the driver? What kind of a truck was it? Did they ever catch the guy?"

In answer to that last question I said, "Too bad the cops never had the interest you guys have. Because the so-called investigating officer acted more like an accomplice."

Here I went into detail about how the ambulance arrived with flashing lights and screaming sirens and left the scene when it was seen that we were only bumped and bruised, and how the fire truck did the same when there was no danger of anything bursting into flames.

Which left us with only the cop. He was a little cop, and after I pointedly offered that this was certainly no offense to the guy whose brother was a cop, I did my best to explain to him and all the others present that what made this particular defective detective seem even smaller was his apathetical approach to the investigation. "I mean, little as he was, the guy was a giant when you put him next to the size of his work ethic. He's in prison now, but if he wasn't, he'd definitely be enshrined somewhere in the Donut Hole Hall of Fame."

"You mean he botched the job so bad they sent him off to jail?"

"How we wished, but no, a little later on he abducted a thirteen-year-old girl. At gunpoint. You wouldn't think he'd need a gun to abduct a thirteen-year-old, but like I said, what we're talking about here is one very little cop."

This brought out more murmurs and more questions. Most of the questions were about the cop and the others were about the Bus. Since, basically, when it pertains to police, the function of the force is to serve as the immune system in the great body of humanity, and since, after all, the members of that particular profession did weed out one of their "own," it seemed unfair to put too much focus on the bad behavior of a single rogue.

So I only answered questions about the Bus.

If they wanted to know where the Bus had been, my answer was most of the states in the US and every province in Canada.

"How about Woodstock," someone said, smiling. "This van *had* to've been at Woodstock."

"I've heard that before, but no, me, I miss out on all the fun. When everyone else was in Woodstock, I was stocking up on wood."

"What?"

"Firewood. This Bus was up in the Yukon back then, and you can't make it through the winter in a place like that without stocking up on firewood. Many have tried. All have failed."

"Well, you didn't miss much," said another guy in answer to the question about Woodstock. "My dad has this friend that was there. He said it was okay at first, but then it got raining and crowded and his girlfriend had to take a dump. And you know what? She did it right there. He put his coat around her and she took this big shit right out there in the field."

"Like I said, I miss out on all the fun."

"No, I'm serious. You should talk to him once. I mean if you want to know what it was really like," and here he quoted his dad's friend on the mud, the drugs, and the music and the garbage, and ended it with, "There were a lot of vans there, just like yours."

"Yeah, back then, these things were all over the place. When I bought this one it was like every third car on the road was a Volkswagen, and every third Volkswagen was a Bus."

"Too bad it's not that way now," someone said, and added with a smile, "Especially with that bunk back there. I'll bet that thing has seen some action."

"Not so loud," I said. "My little bed might hear you and think it has a function other than giving a good night's sleep."

Suddenly everyone was a skeptic, and made their feelings known with a collective groan and a synchronized rolling of the eyes.

I stood my ground. "The truth is still the truth whether it's believed or not."

This brought out louder groans followed by "Now I've heard *everything*."

As long as I was in this deep, I thought why not pile it on. "Drugs too. No drugs in this Bus. It's a total virgin. No sex, no drugs. No radio neither, so no rock 'n' roll. Like I keep saying, I miss out on all the fun."

Still sensing skepticism, I added, "Well, maybe not all the fun, because if you took the alcohol out of all the beer and whisky I tipped back while camped in this thing, and you took that alcohol and put it in the gas tank, then you could probably drive all the way from Cairo, Illinois to Bumph Uque, Egypt, and still have enough left over to toast yourself at the end of the trip."

This primed people to ask how many miles the Bus had actually traveled, and the most accurate answer I could give was, "More than zero and less than a zillion," because years ago I lost track of the number of turnovers on the mile-meter. I also lost count of the number of motors it took to get that far, but while we were talking motors somebody wanted to know what was in there now. With that we went to the back of the Bus where I flipped up the motor lid, and there, bolted to the tranny was a smooth-running, easy-starting little flat-four built by Wolfgang Franke, operating out of Rapid City, South Dakota. Why mention Wolfgang? Cuz he da man. Wolfgang builds you a motor, now that's a motor built to last. "I'll guarantee that motor for 3,000 miles," Wolfgang said, and to the onlookers I said, "I owe Wolfgang a beer. Maybe a dinner *and* a beer, because that motor is now well on its way to two hundred grand and still has good compression in all four cylinders."

To drive home the point I said to those gathered, "Pull that," and aimed a finger at a strange little creature bolted to the back end of the motor. When someone said it looked like a pull-start on a lawn mower, only bigger, I said, "Grab the handle and give it a yank. You'll feel the compression right away."

When everyone just stood there looking like they were waiting for someone else to pull it first, I said, "Wait. Let me flip the key so we can make a good thing even better."

With the key flipped to the "on" position, and with the gearshift in neutral, I said the same words again, only with a walk-through demonstration. "Now," I said, "if you fold your fingers over the cross-handle like such, and if you put one foot on the bumper like this, all you need to do is lean back and pull," which I did, and which brought the motor to life as if it had been waiting all day for a chance to show its stuff.

As quickly as the motor started, its sound was just as quickly drowned out by the reaction of the crowd. First there was this big collective gasp, then people laughing, then clapping and saying things like "No way!" "Awesome!" and "Unreal!"

All the commotion brought more bodies out of the fitness center, which called for another demonstration, but not by me. For the encore, one of the original musclemen stepped up, put his foot on the bumper and successfully started the motor in the exact same manner I done did earlier. He folded his fingers over the cross-handle, put one foot on the bumper, and gave a quick lean back.

And then, to my surprise—but not really—his buddy, the one who had slapped him on the shoulder earlier, took a turn, but kept both feet on the ground and used only one hand to pull the rope. When this effort got the same result as his buddy's two-handed try, he topped off his move with another slap. Just like before, he gave his buddy a quick backhanded slap on the shoulder and followed it with a nod as if to semi-seriously say, "Dude. You need to work on those deltoids."

Ted and Shrader were giving tours up front. While I was flipping the key on and off for those who wanted a try at the rope-start in the back, Ted and Shrader were up in front explaining the setup to anyone who would poke in a head and listen. "Them ain't lawn chairs," I heard Shrader say more than once, "Them's webbed seats." And each time he said it he dissolved into giggles. Ted was having just as much fun as Shrader. He stood in mock pride by the cargo doors, and to anyone new to the scene he made a sweeping gesture to the interior and announced, "This is the Swiss Army Bus." Regarding the stove, he went on with, "Too bad it's too early to drive out on the ice. We could drive out on the lake, fall through the ice—and see that stack pipe there? We could breathe through that until we were found by rescue dogs." All that was delivered deadpan, straight-laced and straight-faced, but every time he looked over at Shrader, his eyes would close and little puffs of mirth would pop out of his nose.

There was more. There was a roll-out sink beneath the bunk. Clips held fishing rods against the ceiling, "So unless you're a fly, they don't get stepped on up there." Racks for skis and snowshoes were bolted high on both interior sides. A long-handled shovel was up there too, because that, along with an ax and a jack, could get you rolling again after being stuck in anything short of quicksand. And let's not forget the toolbox. I showed how it slid out from under a place invisible to the average eye and how it carried a full set of metric tools, because with all things mechanical, physical, and even spiritual, you never know.

We talked for maybe an hour. And listened for most of that time. When they went away and I left too, I brought them all along. Thought can do that. We motored on up by the lakes and the loons and I showed them here what they showed me there. There was new growth all around, on the earth and in the sky, and there it was for all who choose to see. From the fitness center and the hair saloon to a place in the world without such things, there it was, life in a forward climb, moving from what it was to what it is, and moving on to what it is not quite yet.

A Big Dog

*"The smarter you think you are,
the easier it is to do something stupid."*

(Posthumous comment made by Vandalmire "Ras" Putin, who took the term "Practice makes perfect," and applied it to Russian roulette.)

Maybe in times of war. Maybe in a combat situation you'd save your best buddy's life in the same day he saved yours, but in times of peace? And not just ordinary peace either. This was peace times ten. This was peace elevated all the way to the seventh level of tranquility—maybe the eighth—because the north woods will do that. You pick a time like we did, a time when the late winter sun opened up the creeks but still spared the snow, and you arrive early enough in the day to put up a cold-weather camp with the kind of comfort normally found at the beach—here's where you kick back and say ...nothing, because looking out over all in view and seeing only quiet peace and equally quiet perfection makes you your own sweet self want to be quiet too.

This was National Forest country, made of millions of acres of mostly wilderness in that part of the upper Midwest bordering with Canada. You can get lost up there. You can get lost and never be seen again, or you can lose yourself and never see life in the same way again. We chose the latter, again, as we had chosen before, and still before that, although fate and fatal moves would also have a say, because man is a limited creature in his capacity to survive in his surroundings but at the same time unlimited in his capacity to get his dumb ass in a bind.

First we did the shoveling. Winter in the north woods puts a lot of snow on the ground, so to get the Bus onto a flat spot near a creek we had to dig. We were parked on a narrow little road that connected two parallel roads, each one eight miles apart. The parallel roads had high snow banks and low traffic, and our little connecting road had even higher snow banks and even lower traffic—if you can call something like ten cars per week "traffic."

When I said, "We did the shoveling," I didn't mean that literally. "We" were the team of myself and my best buddy Scotty. Scotty and I go way back. We first met at an orphanage. Even though he was very young at the time, the people at the orphanage already had him pegged as one who had potential and as one who had potential problems. On the "potential" side, they said he was very energetic and intelligent, and in the area of "potential problems" I was warned that he would need a lot of patience and maybe even more discipline because "all he wants to do is play," which certainly didn't seem problematic from my point of view because that's about all I wanted to do too, and with that the adoption papers were signed, we walked away, and soon thereafter it was common for the curious to stop us as we traveled and ask, "What kind of a dog is that?" and when my answer was "Part Collie and part German Shepherd" they'd shake their heads and say "He's beautiful," and yes he was, from stem to stern, both inside and out, so that's why I said, "We did the shoveling," because with our kind of connection, when you do things, you never do them alone.

To dig our way into our chosen spot we needed two shovels, because snow, like truth, beauty, and Grandma's bootie, comes in many forms. A pointed spade shovel is best for breaking up and tossing away the big bank of packed and heavy snow thrown off the road by the county plow, and then once you get beyond the tough stuff you can park your spade shovel and work your way through the fluff with one of those light and flat-bladed plastic jobbies specifically made to move at least 1.633 cubic feet of snow with each toss. It was all about efficiency. After all, when a stroke of fate followed by foul weather wants you dead on day five, you want to live the four days you have left as efficiently as possible.

But who'd ever think such a time was waiting? As far as we knew, what was waiting was a fish dinner, for sure, because just up the road and over the hill was a lake so fertile it seemed to have more fish than water. From the standpoint of a worm or a minnow, this place was predatory hell, but if you're like Scotty and me, a couple of guys who think there's nothing better than a fresh-caught plateful of deep-fried crappies and bloogies, then this body of water represents five stars on a scale going from one to four.

Plus it's very lightly fished. You'd think a lake with a piscatorial population rumored to resort to fistfights in a rush to be the next one caught would have anglers coming in from all directions. You'd think they'd have the ice drilled so full of holes as to appear to be imported from Switzerland, but no, once Scotty and I finished with our shoveling, and once we walked over the hill with a hook and line, there on the lake was not a person to be seen.

No way did this go down as disappointing. Unless your head is full of worms and germs, you don't motor up into the north woods to be lonely for the crowds of New York and Chicago, even though once ago both of those mega-centers of crime and culture probably looked very much like the pristine scene that greeted the appreciative eyes of me and Scotty. Too bad, but that's what happens when places are forced to give up all their diversity in favor of only people.

Since value is often determined by rarity, the harder it is to find people, the easier it is to be friends. During our third day camped in the woods, Scotty and I finally found a friend. Right away you could tell he was a friend just by the tone of his voice. "Are you all right?" he called out with curiosity and concern. This from a car window up on the road.

When I responded that, yes, we were fine, he went on to say that he had passed this way earlier and got to wondering if we were maybe stuck and needed some help.

"Well, actually yes, we could use a little help, now that you mention it," I said, and pulled a paper plate out of the Bus. The plate was full of panfish fillets, deep-fried for dinner, and when I took them up to the waiting car I said to the driver, "Think you could help us eat some of these?"

Here he went from an attitude of "willing to help" to one of "eager to help," and laughing, picked off a fillet. We were both laughing by then, and both lunching down on fish, and each time we picked off a fillet I handed one over to Scotty. "Yeah," the guy said in between bites, "I came by here yesterday after work and nobody was around. He smiled then and added with a wink, "Musta been you was out fish'n."

Fishing was a big topic in the north woods. Hunting too, and pretty soon it would be time to tap the trees for maple syrup. My new friend, who introduced himself as Gary, had a stand of maples he tapped and was looking forward to the thaw. "Two more weeks, I figure. Won't be long, maybe two weeks and the sap'll be running. Here," he said, putting a piece of paper in my hand. "This is the address of our sugar camp." And with that he picked off another fillet, nodded in approval, and followed the nod with "I'd say you got some free maple syrup coming."

Rick and Bernie stopped by two days later. They were charged with keeping the roads open and drove a big county plow to prove it. Even though it hadn't snowed for several days, they still made their regular rounds, snow or not, because, "You'd be surprised what we see along these roads."

That was Rick talking, and turning to Bernie he continued with "How many cars do you think we pulled out of the ditch already this year?"

Bernie gave it his best shrug and said, "Can't say for sure, but I wish they'd go in front first. We need to put up a big sign: When sliding off the road, show us your ass and not your face."

"He means those front-wheel-drive cars," Rick said, and went on to say it was hard to make a hookup on those kinds of axles. Since the rear axle is just a straight bar, you can loop the tow strap in one quick stroke, no problem, but you have to be careful how you pull on a front-end drive train. Hook it up wrong or pull too hard and things can get bent. "More than once we couldn't be sure of the hookup and had to radio in for a wrecker. People don't like that because then they have to pay, but you know how it is nowadays. Make one mistake and everybody wants to sue."

That was the small talk. The big talk was bears. Actually bears and wolves. Rick was a bear hunter with six dogs. Now he was down to four. "Wolves got 'em.

Leave your dogs out loose at night and all you're going to find in the morning is turds and fur. Bernie, how many dogs up here got killed so far by wolves?" and then answering his own question said, "Over 200. That's documented cases. Who knows how many go unreported. I didn't report mine. I don't want to be on the list. Even though you get government compensation for your loss, you also get your name on the list and just who do you think they're going to come looking for when some bleeping heart just so happens to find a dead wolf full of bullet holes?"

According to Rick, the dumbest thing the DNR ever did was use the upper Midwest to re-introduce the gray wolf. "Do you know what DNR means? Do Not Resuscitate. I'm serious. Every one of those bozos at the DNR should be looking for a new job. They probably will be, too, once one of their precious wolves kills somebody's kid. Mark my words. Wolves are getting bolder every day. I think they're taking lessons from coyotes. If people think losing a dog or a cow is a big deal today, just wait till tomorrow. Mark my words. It's only a matter of time before some wolf takes down someone's kid."

The whole time Rick was talking Bernie was nodding. Bernie was a deer hunter, not a bear hunter— "Bear meat tastes like shit"—and as a long-time hunter of deer he said he has noticed that the growing wolf population has coincided with a definite difference in the volume of venison running at large. "Last year was the first time I had to buy meat. I mean I've bought meat before, but last year was the first time I actually *had* to buy it." Wolves. Bernie wasn't as passionate in his opinion as Rick, but he was in full agreement. "It's true," he said. "The wolf has no place up here. They're obsolete. If they were brought back because some bureaucrat had a wet dream about how 'noble' they are, then that's just a plain bad idea. They're cold-blooded killers. And if some other bureaucrat did it to restore shall we say, 'the balance of nature,' then why don't they turn 'em loose in the cities?"

Here Rick cut in with, "Goddamn right. Game is scarce up here. The last thing we need is another predator. You can ask anybody who lives up here. But do they ask us? No, they ask some fat-ass sitting behind a desk. Oh wonderful, wonderful! says the fat-ass. Bring the wolves back! Yes! Yes!" Rick took some time to shake his head and roll his eyes and then added, "Well, the way we see it, wolves are a problem, not a solution." Then he smiled and finished with, "But it's a local problem, and when you have a local problem, that's a problem you best handle locally, if you know what I mean." And here Rick drove home his point by drawing back his thumb as if cocking a rifle while at the same time making a very distinct double-click sound with his tongue.

Even though I didn't agree with everything Rick and Bernie said, Rick's thumb action and double-click sound brought back memories of the years I spent living in a wild neighborhood in the Yukon Territory. Others living in the same neighborhood were bears and wolves, and much like any other neighborhood, there were always those neighbors with an inability to behave. I wanted to tell Rick and Bernie about that one particular afternoon when I had to use gunfire to convince a wolf that I was not easy prey, but it was getting late. Bernie looked at his watch and said, "We gotta get going. Us county workers, we have a reputation for always being busy. It goes along with a saying we have: If you want to keep your job, keep busy."

That was the humorous part of the farewell. The ominous part came from Rick. "We have another saying around here," he said, glancing down at Scotty. "And that is: If you want to keep your dog, keep him close."

Keeping Scotty close was no problem. He wasn't like a hound dog. A hound dog picks up a scent and he'll put his nose to the ground and follow it through seven counties. But with Collies and German Shepherds, their main sense is sight. Over the centuries they were bred for herding and guarding, and for that you have to stay close and use your eyes. Scotty's eyes, whenever we were out and about, usually made sure the two of us were never more than a few paces apart, although he could be temporarily distracted by a squirrel or any other potential playmate and in such a situation might stretch those several paces out to a hundred.

So I thanked Rick for his advice, even though upcoming events would make his advice seem less like advice and more like a premonition.

Regardless, Rick and Bernie were right about one thing. It was getting late. When they left, I checked my watch and guessed that, Oh, we had maybe four hours of daylight left.

Ordinarily, instead of checking your watch, a quick glance at the position of the sun will give you a read-out on how much remains of the day, but a thick layer of clouds kept the sun's position in doubt. For the past four days it had been the same. Except for a few hours of sunlight on our day of arrival, the only view you got when looking up was the sight of a thick layer of clouds. The clouds were also low, and dark, and to look at them once gave the impression of imminent precipitation.

But I had looked at them more than once. I had looked at them again and again and for days and days.

And during those days they had stayed the same, so instead of appearing as a threat, their unchanging state gave an impression of stability. That is, their unchanging state gave m*e* an impression of stability. Common sense, on the other hand, would have asked, "When your bladder is full and your back teeth are floating, isn't this a time to squirm?"

But I had plans. And when a person has plans, common sense is often told to wait. This is nature's way of turning people into compost. Nature has long known that man has morphed into a highly-developed form of pathogenic bacteria. And just like the lower forms of pathogenic bacteria, man will expand at the expense of everything else. So, to keep this beast at least semi-manageable, nature gave man confidence in his own stupidity. You see it everywhere, wholesale and retail, and in all shapes, forms, and sizes. It's especially noticeable in those who very confidently confuse love of country with obedience to government, and send their sons to an early grave, but less noticeable is when it's done personally, in private, and usually in a hurry, by you, maybe, but for sure by me, and with me, this time, all it took was a ten-minute plan packed into a nickel's worth of time.

The creek I was camped on had its origin in a lake about half a mile upstream. It was a small creek, but what it lacked in size it made up for in cold, fast-running clarity. A beaver would look at a creek like that and say "Dam," but me, a guy with a taste for trout, would think "Damn, I'll bet this creek has Brookies."

All the evidence pointed to that being a safe bet. My map showed the creek meandering through miles of national forest. There was no human habitation on either bank. Halfway into its journey from the lake to a distant river below, the creek was joined by another creek whose origin was a spring. So you can see why I didn't want the warnings of fate and foul weather to interfere with my plans. Me wanted to scout that creek. That creek had to have holes where the Brookies were hiding. There also had to be undercut banks at the bends, and dips behind rocks, and one of the best times to locate these potential hot spots is in late winter, since this is a time of no bugs and no leaves, and with that you can make the most of a scouting trip, not having to contend with biting bugs and blinding vegetation.

Scotty was ready. He had that look. When it's just you and your dog—that is, when much of your time is spent in places where others are seldom seen, you and your dog acquire a special form of communication. Normally when people think of communication, they think telephones, TV, e-mail, gang slang, profanity, and, if you really want to get kinky, a post card or a letter. All words and sounds. But. There's more. There's much more, and most of it comes from the realm of movement and gesture generated by thought. However, that particular realm of thought is often over-emphasized and hyper-dramatized by those members of society falling under the heading of "crazy people." Nobody wants to be lumped in with crazy people—unless it's an election year—even though some of the stigma has been lifted by upgrading the term "crazy people" to "People of Crazy." (It be all about changing the name to keep ahead of the meaning.) Knowing that, us normal folks tend to be skeptical of anything "too subtle" while we're circulating in public. Privately though, we know this subtle form of communication to be mysteriously real—especially when it performs in a pinch, which is why I mentioned that Scotty had "that look," a look presently of one who knew we were about to move. He saw me getting ready. He watched as I put the ax and the saw inside the Bus, and when I pulled out my hat and mittens, he put on that happy look that said, "Great! Time to hit the trail!"

Soon though, he would put on another look. It would be a look of exasperation and impatience, it would be directed at me, and it would perform in the ultimate pinch.

I should have known. I mean, along with taking my hat and mittens out of the Bus, I should have known to also take my survival pack. This was a small packsack with all the necessary items to keep a guy alive overnight while stuck in the sticks. The main pouch held a small hatchet and a mini-saw, a saucepan, and a plastic tarp just big enough to function as an emergency lean-to. In one of the side pockets were matches and a compass, while the pocket on the other side held luxury items. Ordinarily you wouldn't think of a simple chocolate bar and a package of dry soup mix as luxury items, but then ordinarily you don't have to do an overnighter while stuck in a Cold Wet Spot in the Deep Dark Woods already. Luxury. It has less to do with cognac and caviar and has more to do with circumstances and effect.

But none of this was thought of at the time. At the time, all I wanted to do was take a quick run at the creek. I had it all figured out. Travel light, move fast. It was all about using the amount of daylight left as efficiently as possible, and every idiot knows that efficiency is synonymous with speed, and if you want to be speedy you need to travel light, even if it puts you in the dark.

So I left the packsack hanging on a hook inside the Bus.

I also left my snowshoes in the Bus, because, as previously mentioned, "I had it all figured out"—which meant my scouting trip would not be done in the usual fashion. No, instead of strapping on the snowshoes and mushing downcreek directly from the Bus, I'd approach the creek by walking a loop. I'd follow the gravel road, and even though my map showed the road angling away from the creek, it also showed a connection with a trail, and showed the trail as a path winding through the woods to a place where it ended at the creek. This looping approach made sense in more than one way. Not only would it put me in the most likely place of Brookies—where the main flow of the creek was joined by the aforementioned tributary originating in a spring—but it would put me there without having to use the snowshoes. I knew from past visits to this area that the trail intersecting the road was occasionally used by snowmobiles, so that meant the snow on the trail would be packed, and that meant the packed snow would let a person hike his way to the creek without having to break trail with a bunch of long and clumsy snowshoes. It was all about efficiency.

Speed, too. Without the snowshoes and without the crushing burden of the seven-pound survival pack, I was able to trot my way to where the trail crossed the road, and once there, sure enough, the snow was packed and made the trail easy on the feet.

Speed and efficiency. Efficiency and speed. Any way you looked at it, you had to praise the combination. I was moving along quickly, and in big easy strides. It was almost like a very confident and well-planned attack.

And here history might have added, "Isn't this the same confidence the Germans showed when they rode into Russia back in World War Two?"

The first sign of trouble came at a hill. The trail took a turn from east-west and headed north. Soon it came to a hill, and as I made the climb, one foot broke through the crust and had my left leg sinking in past the knee. So what, I thought. This kind of thing always takes you by surprise, but it's really to be expected, especially on a south-facing slope where the crust is directly exposed to the rays of the sun. Sure, the sky was cloudy, and had been for days, but some of the sun has no respect for clouds. Some of its rays pass right on through, bringing heat. It's an invisible heat, a heat that warms skin and softens snow, and even though this kind of heat at this time of year gave hope to all who wish for spring, all it did for me was have me feeling like a fish—a flounder, to be exact, because whenever the trail exposed itself to the southern sky, I could expect the weakened crust to break under my feet, and whenever that happened in a serial fashion, what I did was "flounder."

I thought of going back. Mostly because I could feel it getting warmer. At first I thought the feeling of warmth was due to my own struggle. Extra muscle effort will do that. But then my feet started breaking through the crust in areas of the trail normally shaded. So that meant the same sun that weakened the crust in the exposed areas was also warming the air, and that added warmth was putting a wrinkle in my plans to travel with speed and efficiency.

What ironed it out was the memory of my map. My map was back at the Bus, but as I pictured its delineation of the trail and put that next to my position in the woods, it seemed like I was much more than halfway to the creek, and when you're more than halfway to anything the natural tendency of a human bean is to keep going, even if history and past personal experience are whispering "Go back."

As it turned out, I was even closer to the creek than originally thought. Making the situation better yet was the lack of snow along its banks. Mostly grass greeted me. It was that thigh-high kind of typical creek-bottom grass that falls down flat after the first frost and makes easy walking even easier. Which turned out to be good, bad, and then good again.

The good was that now I could poke along the creek unimpeded. In this particular section I could look for holes and eddies, undercut banks, and dips behind rocks. I could do all that without breaking through crusts, wading through the soft stuff, and making a moron out of myself by punctuating the struggle with attendant bursts of profanity.

The bad part came when Scotty followed my example. He nosed along behind me, pawing through the grass and sniffing for mice, and somewhere in the middle of it all he decided to take a drink. He decided to take a drink from the creek. You couldn't blame him for his choice of water, cold and clear as it was, but you had to fault him for the place he chose to belly up to the bar. It was a trap. The bank of the creek, instead of dropping off abruptly, angled down to the water at about the same slope you see on a playground slide.

Scotty found out it was a one-way slide when he stepped to the edge of the creek. Here his front feet quickly sank into a silted area, while his back feet both tangled in the interwoven mat of fallen grass. There was no escape. The angle of the bank and the gripping of the grass were both too much to let him back up, and the soft silt in the creek bed held his front feet as if he were mired in quicksand.

And just like in quicksand, the harder he tried to get out the quicker it pulled him down.

Lucky for him we were close. We being close, there was time for me to rush back before his efforts to be free had him sinking in over his head. As it was, I came in quickly from behind, took hold of his tail with both hands, and leaned back with enough of a yank to haul his butt back up to a safe part of the bank.

When he looked at me as if the pulling of his tail was more of an insult than a rescue, I said right out loud, "Lucky for you we were close."

Little did I know that I should have said, "Lucky for *us* we were close," because the day wasn't done with us yet. A little further on, when the day was dying, it would serve up a particular form of fate to imply that the day didn't want to die alone.

But the future I held in my head was different than the one soon to happen. My future was not about the closing day, but about opening day. That is, opening day of trout season. Already the creek was looking like a likely winner, and the farther I poked along its banks the more it seemed to agree. The habitat was all there, and the excitement of finding it kept me on the scene longer than planned, especially when having to thread my way through a tangle of tag alders. Tag alders are bushes posing as little trees that absolutely refuse to grow straight up. Their wrist-size trunks and branches shoot out from central clumps at shallow angles and interlock with other alders growing from clumps nearby. Picking your way through stuff like that takes time. You go through a lot of pushing and ducking and lifting, and lifting and ducking and pushing, and after only a few minutes of that you understand why some call tag alders the Devil's Turnstiles.

Still, they had their upside. In places where they overspread the creek and connected with others of the same kind on the opposite bank, they added to the habitat. In the warmer months the canopy kept the water cold and was a valuable source of dropping insects for the trout waiting below.

When the tag alders gave way to another open area, I realized there was no point in going any farther. The creek had revealed itself to be a solid source of future fish fries, and barring any "progress" brought in by Big Oil, Big Frack, or Big Mac, the creek was likely to stay that way at least as long as it takes water to wear away rocks.

That much was assumed.

What was also assumed was that picking our way back through the tag alders might not be necessary. Instead of retracing our steps back through a bunch of tangled branches that didn't seem to want us there anyway, I thought why not double-team the situation by avoiding the tag alders and taking a shortcut at the same time. The more I looked around and thought about our position, the more I thought why not. Sure. I kind of knew where we were. Even without the sun, and even without the map and compass I was pretty sure of our current position in relation to the trail. That is, Scotty and I took a right turn when we left the trail to follow the creek, and while we were following the creek, it too, seemed to bear to the right. So that meant in all likelihood that this section of the creek was running parallel with the trail—which seemed to coincide with my memory of the map.

With that much evidence, it seemed foolish to retrace our steps. Especially since we'd spent more time on the creek than planned. Right. Daylight was starting to fade, and whatever Scotty and I could do to get back to the Bus before dark made perfect sense.

So I cut away from the creek. It was all about geometry. Geometry tells us that girls have the curves and guys have the angles, but even more important than that it says the shortest distance between two points is a straight line. So I cut straight away from the creek and took a perpendicular line up into the woods. The plan was to hook up with the trail in the same way as cutting through the middle of the block will hook you up with a parallel street, thereby saving you the time and trouble of hoofing and huffing your way back around.

And once into the woods, it wasn't just geometry that seemed to be on our side. After two steps into the timber we lucked onto an animal path that made its way through the trees as if it were part of our own plan.

But there was more. "Look here, Scotty," I said, pointing to a paw print in the snow. "One of your ancestral brothers." The print was too big to be from a fox or a coyote, and was not put down by some passing St. Bernard or Great Dane. That much was certain when a little farther on here in the trail is this big turd. It was of recent manufacture and revealed itself as an anal sculpture typical of a wolf in that it had the size of a stretched and stuffed bratwurst and was mostly made of hair. Say what you want about wolves, but save a moment of silence for one of God's more resourceful creatures who has no qualms about filling a requirement in dietary fiber by fistfuls and mouthfuls of nothing but hair.

We motored on. The trail petered out, but in its place appeared a guiding line of spruce trees. Once again, it was almost like part of our own plan. The spruce trees, with their dark needles and dense branches, grew in a line side by side with a stand of pines. Pines have lighter needles and fewer branches. So if the question was, "What's the best way to keep a straight line while walking this area of the woods?" the obvious answer was, "Stay on the line between the spruce and the pines."

We walked among the spruce trees. The branches of the spruce, being dense, kept most of the snow in the crowns so what little fell to the ground was only intermittent, icy, and covered with fallen needles—much easier to walk through than the knee-deep snow below the pines. Soon the line between the two different types of trees became less distinct, and finally gave way to a dense forest of nothing but spruce.

Ordinarily this would have been a good place and a good time to turn back. Let me rephrase that. Ordinarily this would have been an *ex*cellent place and time to turn back because suddenly the whole scene fell under the heading of "iffy." Iffy in the sense that the decision to take the shortcut was based solely on assumptions. And chief among the assumptions was the belief that my chosen line of travel would intersect the trail. Oh? Then why weren't we there yet? Something that was supposed to be a shortcut was taking too much time already. The feeling was that the time spent on the "shortcut" should have already brought us to the trail. Which basically boiled down to two alternative questions: "Are we going in the wrong direction?" or "Are we almost there?"

It seemed I should give this some thought. Before making a decision to feel my way forward into the trackless spruce forest or to very discretely take the well-marked path back the way we came, it seemed I should pause for a moment and give the situation some serious thought.

But there was no time for that. The day was almost done. The fading light seemed to say, "Either go back or go forward, but above all, go."

So why not do both? Or at least have "both" as an option. That is, first feel my way forward for maybe five more minutes. Take about five minutes of the remaining day to pick a straight line forward through the spruce trees. Mark the way by breaking the occasional branch and having it hang down conspicuously. It's an old Indian trick. You break a dead and brittle lower branch and have it hang down in kind of a flag fashion. Keep doing that as you walk along so that when you turn around and walk back the hanging branches point the way.

Not a bad idea because spruce trees, with their dark needles, soak up a lot of light. This particular grove was tightly packed too, so even before my five minutes were up, the trees were presenting problems. Unseen branches were poking me in the face, and the density of it all made trying to travel in a straight line virtually impossible.

What kept me going was an appearance of light up ahead. Somewhere in the middle of breaking branches, pushing ahead, and having second thoughts about the whole process, I looked ahead and saw what seemed to be a line of light crossing from left to right.

Yes! I thought. The trail! Of course! The gap in the trees over the trail let in light from overhead, and that light reflecting off the snow was in direct contrast to the dark and snowless ground at the base of the spruce trees.

So flush with a sense of discovery and relief, I plunged forward until coming to the line of light, and sure enough, just as assumed, the light was the result of overhead illumination reflecting from snow on a trail.

That was the good news.

The bad news was there were no footprints in the snow. No footprints and no paw prints to show that some guy (me) and his dog (Scotty) had passed this way earlier.

Immediately and dreadfully I realized my mistake. Not only were we in the wrong place, but in my hurry to get to it, I failed to break off the tell-tale branches to mark our way back out.

Making it worse was the weather. The clouds that had held off for four days were now dropping snow mixed with rain. This went unnoticed while I was under the dense crowns of the spruce forest, but now that I was out in the open, the falling rain and snow was both seen and felt.

Great, I thought. It's late, I'm getting wet, and my sense of direction has dissolved into sensible doubt.

Still, it wasn't the end of the world. Maybe it was the beginning, because trails in the north woods always lead to something. Most of them were cuts made long ago by loggers to reach the timber, and by the looks of the track I was on, this was one of them. Maybe it was even one that crossed our path while Scotty and I were mushing our way to the creek. This was a very distinct possibility, since I had seen remnants of more than one old logging road crossing our path as we came in.

With that in mind, I decided to take a chance. Instead of immediately turning back and locating the line of broken branches while it was still light enough to see, I decided to pursue the possibility that the trail I was on did in fact intersect the trail I wanted to be on.

First I went to the left. And quickly. This was no time to be screwing around. It was late and I was getting wet from both ends. Falling rain and snow was seeping through my clothes from the top, and marching through the wet and clinging snow on the trail was doing the same thing from my knees on down.

Scotty seemed to sense that something was wrong. Instead of stringing along in his "typical Scotty" fashion—which meant bouncing to the left and then to the right, and running ahead and falling back—instead of that he stuck close to my side and plodded along like a soldier in the middle of a long and serious march. There was something about that. There was also something about the position of his ears. Normally his ears stook straight up, German Shepherd style, but as if to coincide with his plodding, now they were held halfway down to the side and laid slightly back. I wanted to think his ears and his walk were the result of his reaction to the rain and snow, but what got me to thinking otherwise was the sight of the end of the trail. The trail simply stopped. It didn't peter out or suffer blockage by a bunch of fallen trees and branches. It simply ended and that was that.

In a different situation and under different conditions I would have said, "So what." If it wasn't late and I wasn't wet I would have looked at Scotty and said, "Oh well, let's just turn around and see if we can make our connection by traveling the other way."

We did turn around and we did travel the other way, but there was nothing "Oh well" about it. "Oh shit," maybe, but not Oh well, because I was about to lose one of my senses. Soon I would be somewhat sightless. The fading light would take away most of my sense of seeing and leave me with a sense of loss, so I very quickly did a turnabout and pushed in the other direction. Scotty followed me as before. We passed the place where we entered the trail and kept right on going. Fast. It was my last bid to make use of the "shortcut." Right. Make a quick move in the other direction in hopes of hooking up with the trail. It was all about speed. Move fast in the time left. Go, and keep going until either coming to the trail or coming to the conclusion that this is not the way to the trail.

Right away there were problems. Whoever made the logging road made it far enough in the past to allow new growth to sprout up where it was able. When we marched over a rise and down into a dip, what should we see but a thick stand of new spruce. Head-high, it was filling the narrow little road as if put there to function as some sort of a deep and prickly fence.

A wet fence too, because anyone pushing through that was only going to get heap big soaked and heap big cold, since the density of the branches held bushels and buckets of saturated snow, all of it poised to drop at the slightest touch.

So I looped to the right and made my way through a stand of poplars, all the while keeping my eye to the left. The idea was to make a detour around that part of the road plugged by the thick stand of saturated spruce trees.

Once back on track there was no relief. Not only was I starting to feel physically fatigued from all the leg-lifting needed to get one foot in front of the other in deep snow, but then something presented itself to make my mind even more tired than my body.

We came to a fork in the road.

Now we had two options in a situation where there was hardly time for one.

Here's where I realized that the fork was not really a fork at all but in fact a dead end. No way would time allow a continuing pursuit of "the shortcut."

I only hoped it wasn't too late to find "the long way back."

Finding the first part was easy. Our tracks in the snow saw to that. But when we came to the part where we left the forest and entered the logging road, all bets were off. There were no tracks to guide us through the trees, and even if there were, it was almost too dark to see them anyway.

It wasn't getting any lighter just standing there, so I very carefully stepped off the road and into the trees. The idea was to take a few steps at a time and then stop. Take a few straight steps into the trees and stop, and then look back and look forward. Be sure of direction. Keep focused on going straight. Since we came straight to the logging road from the woods, going straight into the woods from the logging road would likely put us in the place where I broke off the last branch.

Once again, everything was taking too long. Way beyond the time it was supposed to take to connect with the broken branches, I was still searching. Groping, really, since light among the spruce trees was by then almost nonexistent.

Not wanting to get "lost," I figured I'd take a few more forward steps before turning back. Just one last-ditch effort to hook up with the broken branches. Then, if they still failed to appear, I'd thread my way back to the logging road and try to figure out what to do from there.

Well, a few more steps and wouldn't you know. We're on the trail! I mean, *we are on the trail!*

I couldn't believe it but there it was. Scotty and I stepped out of the woods and into the open and what do we see? Footprints and paw prints on a trail. Unreal. I smiled and shook my head and said right out loud, "How's that for luck?"

Turned out it was bad luck. A closer look revealed "the trail" to be the logging road. We were back on the logging road. The "straight line" taken into the woods ended up as a circle and had us stepping back onto the logging road in almost the exact same spot where we left it.

Here's where I almost felt as if someone was watching. It was as if an invisible someone had been watching since even before we left the Bus, and each time I made a move, this particular someone would cringe with the knowledge that it was the wrong move—wrong to start late, wrong to leave without the survival pack, wrong to pursue a "shortcut," wrong, wrong, wrong. Dumb too, and not just dumb in my decisions but doubly dumb not to question them. At this, I could almost hear the other someone say with a sigh, "Now you're lost. You're lost and the only direction you can be sure of is down."

Down. That gave me an idea. Travel down. Never mind a bunch of trails and logging roads. Simply travel down. Take a lesson from the raindrops and follow the contour of the land. Follow it downward until coming to the creek. Of course. The creek. The creek had to be lying in the low point of the surrounding territory, so all Scotty and I had to do was follow the downward contour of the land. Soon we'd be at the creek, and from there we'd easily know the way out.

"C'mon, Scotty," I said. "We're going home."

My optimistic outlook was dampened a bit when we came to what seemed like a solid wall of tag alders. We were moving along, keeping to the low areas and threading our way through mostly birches and poplars, and pretty soon there's an occasional clump of tag alders. Then a few steps beyond that and here's this solid wall of the same stuff.

To keep from thinking of them as a barrier, I thought back to when we were scouting out the creek. Down along the creek, tag alders were growing all over the place. It's what they do. They don't grow up in the hills. They like the low country, so with that in mind I began to push my way through their interlocking branches, thinking, "Well, didn't we do this before? I mean, when we scouted the creek didn't we pick our way through a bunch of this before? Of course we did, and maybe we're back in the same place. Maybe two more steps are all we need to put us in sight of the running water."

There was running water after two more steps but it wasn't water running in the creek. It was water running down the back of my neck. With wet snow collecting on every surface, you couldn't push up or climb over an alder branch without a lot of slop shaking loose. Even when I kicked the branches before squeezing through— even then they held enough water in reserve to make me realize that further travel through this kind of stuff was not only futile but dangerous.

Scotty was already out. When I made my way back to the birches and poplars, there he was, sitting in the snow and looking at me. Something about his look got me thinking. Wait a minute, I thought. He didn't follow me into the tag alders. The whole time I was floundering around in there he waited at the spot where I went in. He always follows me wherever I go, but this time he didn't do that. Instead, he sat down and waited at the place where I went in.

There was no time to think about what that meant. Before it got any darker I needed to find an overlook. I needed to climb to a ridge or some other high spot in order to get an overhead view of the land below. It was my last chance to find out if the creek did in fact run through the tag alders.

When I found that it didn't—when I reached a suitable overlook and saw nothing but a swampy bowl surrounded by ridges—here's where I looked at Scotty and understood why he didn't follow me into the tag alders.

He did follow me to the next ridge. And the next ridge after that. It was my last option. To keep from traveling in circles, I'd stand on one ridge and take a sight line to the next. Without a compass and without the guiding presence of the sun or the North Star, following a sight line from ridge to ridge was about the only way I could be sure of a constant direction. Not that I knew *which* direction. It's just that by maintaining a straight line instead of walking in circles a person has a better chance of hooking up with something other than a re-visitation of past mistakes—if you know what I mean.

Still, straight line or not, soon we're back on the logging road. By then my mind was in a state where nothing surprised me anymore. Mushing too long through too deep of snow had me too sore and too tired to show a lot of concern that all my efforts brought us back to a place where a mistake was made. It was time to give up the quest. At least for now. For now, I'd have to unsling the survival pack and settle in for the night.

Right. Find a big lone spruce and pull out the tarp and pitch it under the spreading branches. In a survival situation, shelter is always the first priority. Get that tarp set up at an angle under the dense branches of a big spruce and then go out and gather up lots and lots of firewood. No skimping. Make sure to bring back enough to last the whole night because once you get a fire going and once you settle in, what you want to do is put all your focus on drying out and keeping warm. I couldn't emphasize that enough. I was cold and wet and tired and had ten hours of darkness and drizzle to look forward to. Wet and tired as I was, the chances of making it through that kind of a night without protection were next to nothing, so the thing to do was to set up a survival camp and wait for the next day.

Except everything I needed to set up such a camp was back at the Bus.

It wasn't just me who was screwed. There was also Scotty. When I thought back to what Rick and Bernie said about wolves and put that next to the paw print and the anal evidence we saw back on the animal path, Scotty's odds for survival were in deep shit. Me, I'd gradually slip into a state of exhaustion and from there simply sit down or fall down and not get back up. Scotty, though, wouldn't have it so easy. Being too loyal for his own good, he'd sit next to my dead body until the meat fell off the bones. But long before that the wolves would move in and make a meal of him. Maybe they were moving in already. I thought back to a time when I was waiting out a snowstorm in the Yukon, wrapped in a blanket and huddled next to a fire and thinking I was alone when the whole time I was there a wolf was sitting thirty feet behind me, waiting. The memory of discovering those paw prints and that butt print when the storm stopped made me feel I owed

Scotty an apology. We were back on the logging road, once again moving towards the place where we entered it from the woods, and while we were walking I told Scotty I was sorry. Already I was feeling that certain kind of resignation that comes with the approach of extreme fatigue, so before I got too dippy to say what needed to be said, I stopped and reached down and scratched his ears. "That's a nice set of ears you got there," I said. "They should be hearing the crackle of a fire right about now, but the thing is, I blew it." I went on to explain that even though it wasn't his fault that I left the survival pack back at the Bus, he was still going to be the one to suffer the most from it not being here. "Me, I'll just get tired and fall down and fade away, but if I know wolves, they'll move in when I'm gone and tear you apart from six different directions." At that I started walking again, because the effort of walking was the only thing keeping the chills away, and when I thought about the chills I explained to Scotty how things were supposed to be. "We're supposed to be under the tarp and next to the fire. We're supposed to be warming up and drying off. Making plans, too. We should be making the kind of plans needed to find our way out in the morning." Right. No more guesswork. Come the morning, we needed to take the compass to the "hub" and from there "spoke out." I explained to Scotty that "the hub" was where we first stepped onto the logging road, and that "spoking out" meant using the compass to walk out in straight lines from that central place. Walk out as if we were spokes on a wheel for, oh, ten minutes, and if ten minutes of walking didn't bump us up against the line of broken branches or something else just as familiar, then we'd simply go straight back to the hub and spoke out in a slightly different direction. Since the spot where we first stepped onto the logging road was absolutely positively no more than a ten-minute walk to something we'd recognize as familiar, all we had to do was keep walking out in straight lines from the hub. Use the compass to take a slightly different tack each time. Go out, come back, go out, come back. It always worked before. All you need is time, light, and life.

But when we finally got to the place where we first stepped onto the logging road—the place that was supposed to be the hub—time, light, and life were in short supply, and in response to that all I could do was simply stand under the dripping sky and prepare to accept the inevitable.

Even though I was trying not to let it show, Scotty must have picked up on my dejected state of mind. Instead of staying at my side, he stepped to the edge of the woods, took a sniff, and suddenly wheeled around and looked at me. When all I did was merely look back, he bounced forward in one big leap and came down hard on his front feet, all the while still looking at me, and in the next instant wheeled back around and disappeared into the woods.

Ordinarily, this behavior would be seen as a playful move. Dogs do that when they play. They bounce and frolic. But the way he came down on those front feet and the way he looked at me when he did it struck me as an expression of total exasperation. I mean it actually struck me in the form of a definite mental impact. With his look and with his moves he put his own thought in my head, and in the shock of feeling exactly what he felt at that particular moment I instantly knew what he wanted. He wanted me to follow him. He made me very abruptly aware that he was plain sick and tired of wandering around in bad weather and wanted me to follow him back home.

But the problem was, he was in the wrong side of the

woods. I was totally certain that the way we first stepped onto the logging road was from the other side of the trail.

While I stood there with my utter certainty dissolving into a state of certainly confused, Scotty reappeared. He stuck his head out from among the trees and gave me that same look he gave me before. No two ways about it. He wanted me to follow him.

When he wheeled around again, I was right behind. It was dark in the woods, but I could see the movement of his white feet. He's tawny on top, which made his body all but invisible in the non-light of the forest, but all four of his feet are white and were guiding me along like little signal flags. If I got hung up in branches or if he got too far ahead, he'd come back, bump me with his nose, and once again point the way.

I followed him on faith. I followed him completely on faith until we came to a part of the woods where the spruce trees bordered up on a line of pines. Here the ground was snowy and showed footprints. A closer look showed paw prints too, and all pointing in the direction from where we came.

So you know what that meant. That meant I wasn't just following Scotty on faith anymore. Now I was following him on faith plus trust and gratitude, because those footprints and paw prints were ours, and as we followed them they led us not in circles but in a straight line between the spruce and the pines, on to the animal path, past the turd, and down to the creek.

We limped in way late. When I opened up the Bus, Scotty pulled himself up onto the bunk, plunked down, and was immediately asleep. I made a fire in the stove, slumped in the chair, and as the flames put out a glow beyond what math can measure, I looked at Scotty and smiled. Now there was a guy who knew how to relax. After putting in a hard day, he knew exactly how to end it. There he was, with his head on the pillow and stretched out at full length, which means he took up the whole bed, since he measures more than six feet from the tip of his nose to the end of his tail. A big dog. You can't have a dog that big in bed with you when the bed itself is only two feet wide and six feet long, so there was another bed, a smaller one, at the back of the Bus. That bed was his, but he preferred mine because it was the only soft spot in the Bus where he could sleep stretched out. He took all his day naps on my bed, but whenever night came knocking and it was time for me to hit the hay, I'd rouse him and say, "Move over, Rover," and that's what he'd do. He'd move over, I'd move in, and there we'd be, burrowed into our respective beds, comfy, cozy, just two of nature's cuddly little critters in a blissful state of beddy-bye.

 But when I thought of how much he preferred my bed over his, and when I put that next to the sight of his white feet taking us both back to the Bus, I let him spend the night right where he was, and not just because I was too tired to move.

You Can't Stop Progress

*"They locked the gate from Here to There with
faulty signs that said beware."*

(Junkyard Kipling)

  You could see it from the road. It rose above the fence and was framed against the sky like an eyesore or an opportunity, depending on your own horizon. Like a mind, it defied any sense of physical measure, but like a brain it had a definite size and a typical shape. As a mind, it took in all that was offered, and as a brain, whatever was offered was processed. Noise was part of the process—loud noise, sharp, clanking, and booming noise, with most of it made by the grinding efforts of heavy machines moving mountains of metal. This was a scrap yard, a wrecking yard by another name, but if a land claimed by a king can be called a kingdom, then so can a land claimed by wreckage be called a wreckdom.

  The function of the wreckdom was to take in metal that was no longer able to serve. In the Darwinian School of Economics, an object must be discarded when all the use has been squeezed out. Much like a form of digestion, objects are consumed by the economy, nourishing it as they pass through, and when they have exhausted their capacity to serve, what remains ends up in the wreckdom.

  Aluminum, copper, brass, stained and stainless steel—it was all there, piles and piles in the form of pipes, wire, whole cars, washers, driers, tractors, sheet metal, cast, forged, and machined pieces of other metal, and thousands and thousands of things that used to have use. Maybe a part was broken. Maybe a vital part or even one that was trivial slipped a cog and was condemned to be tossed. Just as often something suffered the same fate because it belonged to the wrong date. You hear, "It's too yesterday, and this is today." People are funny that way. They forget that yesterday, today, and tomorrow keep changing places. Ask any dealer in antiques. Listen to the whisper of history.

Meanwhile, whole and broken objects were trucked in from all directions—a flatbed loaded with junk cars, a pickup with a water heater and a tired old stove, and even some guy in a Buick with a trunk full of aluminum cans. Here their cargo was weighed and the people were paid, usually pennies to the pound for stuff sold as scrap, the last chance for the consumer to squeeze just a little more use out of something no longer able to serve.

However, there was a sign-up sheet. I first heard of the sign-up sheet from Ejo Hartmann, long before he died. Ejo lived more than a hundred years if you count life in events, but only half of that if you go strictly by the numbers. There was no stopping him. He was the Redneck Renaissance Man. He knew which bars had the coldest beer, and which lures caught the biggest bass. He hunted too. He got a lot of satisfaction from hunting deer, ducks, and other wild edibles but had his best moments when tracking down bullies. Bullies? "Yeah," Ejo once said, "I found another one last week," and went on to say that while he was sipping a beer in a bar in a visit to his old home town, "There he was, that sonofabitch," meaning which "that sonofabitch" was another sonofabitch in a long line of sonofabitches who once ago liked to pick on The Little Fat Kid, The Little Fat Kid being Ejo himself back as a pre-adolescent child. At this point your question would be "Oh? A little fat kid?" not believing that the tough and muscular Ejo had a former life as a soft and pudgy little punching bag. And not only that, but he was also on the receiving end of atrocities even more galling. But let him tell it. "Anyway, there he was, that sonofabitch, the same one who shoved the dogshit down the back of my neck. I knew I'd find him someday, and goddammit, this was the day." Did you introduce yourself as his former victim and offer to settle things outside? "Too easy. Besides, I didn't want an audience so I waited till he went for a piss." Ejo followed him into the john and when they were both inside he locked the door and very calmly asked the guy if he knew who he was. "No," was the answer he got, so Ejo, to refresh the guy's memory, reached into his own memory and reeled off a long list of abuses foisted upon him as a child. Many of them were perpetrated by the guy standing in front of him, and as Ejo reeled them off he punched the guy in the mouth, slugged him in the stomach, and bounced him from the wall to the sink and over to the toilet. After a knee in the nuts had the poor bugger on the floor by the toilet, Ejo leaned down and once again posed his original question, even though by the look on the guy's face Ejo knew he already had his answer. "But you know what? He took it like a man. Instead of hollering for help or crying for his mama, you know what he did? He wiped the blood off his mouth and said, Well, if I didn't have a reason to respect you then, I sure do now. He actually said that. I swear to God. You don't see that anymore so I help him up and we shake hands. Can you believe it? I hated that fucker. Now we're friends. I tell him, you need anything fixed, stop by my shop."

In the Yellow Pages his shop was listed under

"Auto Repair" but Ejo would take on anything mechanical, plus put on the paint and the polish. Welding? Stick, acetylene, tig and mig. Hoists, lifts, rolling boxes of tools, creepers, and yes, barbells and a heavy bag. The barbells were parked between the air compressor and the workbench and got a lot of use whenever a job would bog down with the kind of frustrations that made Ejo want to throw the tools more than he wanted to work with them. Nearby, the heavy bag was hanging from a long rope tied to a high rafter. "You want it to swing," Ejo would say, explaining that a swinging bag kept a puncher on his toes, and drove home the point by backing up the bag with a quick combination of bare-knuckled jabs and hooks, side-stepping when the bag did a fast-forward, and then on the backswing there was Ejo standing firm and bringing the quickly approaching bag to a dead stop with a stiff right and a high kick.

Kicks of a different order came after hours. Business hours were from 7 to 3, Monday through Friday, and on Friday from 3 till who-knows-when the place turned into a tavern, an eatery, and a crapshoot. Fellow mechanics, frequent customers, and old and new friends would start rolling in a little before "quitting time" with beer, wild game in and out of season, fish, and more beer. Ejo would be putting away the tools by then, and plugging in the deep fryer or firing up the grill—maybe both. He'd gather up the checks, bills, receipts, and the mail from a countertop that functioned as part of the business during the day, and once all that stuff was stuffed in a drawer, the countertop became a bar. Was it legal? If not, then the law most frequently broken was the law of averages, with Ejo the proudest of the perpetrators. "I swear to God," you often heard him say, "I can make the dice mine." He was talking about his talent for rolling the bones to land sunny-side up, and he'd show it whenever a square of plywood was set up on the end of the bar to function as the backboard in a crapshoot. Many were the skeptics. They came in flush and went out broke, and to watch the way it went down was to see Ejo take command of the scene the same way a preacher takes command of a tent revival, blowing on the dice as if to bring them alive, talking to them each with love, and holding them as if his hand had the power of a womb, all the while pacing, chanting, and working himself into a state of shaking connection with whatever it took to make the little black dots add up to winners on the little white squares. To watch the performance and to dismiss the result left the unbelievers sorry that they put their money where their mouth was. "I can't explain it," Ejo often said, "but I can make it happen. Once I get going and once I get this big headache coming up the back of my neck, then I know I'm in the zone."

Even if you didn't believe in his talent for holding his own "in the zone," you had to believe in his generosity. If some poor sap lost all his money to Ejo in a crapshoot—money that was supposed to pay for repairs on a car, Ejo would often write it off, saying he needed a tax break anyway, or else would let the guy use his shop and his tools to fix the thing his very own self. "You come in too," he said to me when he heard I was replacing the rusty body panels on my Bus. "Bring it in tomorrow and we'll set you up. Get some pop rivets. I have the gun. You can pay me in beer."

When I motored in the next morning with the Bus, a box of pop rivets, and a full case of his favorite brew, he was good for his word, and then some, by saying, "Any sheet metal you need you can pick off the rack in the back. You want lastability, so I'd go with the galvanized."

"Thanks," I said, "but how about this?" Here I reached into the Bus and pulled out a pre-cut and bent-to-order piece of 16-gauge sheet stainless.

"Did you buy that from a metal shop?"

"They wouldn't let me leave without paying."

"No, really. You bought that from a sheet metal shop and you still have both of your arms and both of your legs?"

"I'm not saying it was cheap. All I'm saying is, once it's on there, it snot coming off."

"Quick," he said. "Get in the truck."

"I'm not taking it back, if that's what you mean."

"No, no, leave it. It's poison. Put it down and get in the truck. Quick. We have to go right now."

He was already on his way out, so there was nothing to do but follow him. And once in the truck he drove with the same aura of urgency he showed in the shop. He pulled the shift lever quickly into gear, hurried around the corner, and cursed and fumed whenever we got stuck behind somebody who was driving the legal speed limit.

"Just where the hell are we going?" was not going to get a factual answer out of Ejo. No way, so I took note of our direction and said, "You having a baby?"

For that I got "his look." It's the same look you see on the face of a parent just before he swats a lippy child.

"Well," I said, giving it my best shrug, "we're racing toward the hospital and you're having a hard time keeping the pedal off the metal, so I thought maybe your water broke back there and now you need some medical help to pull out Little Junior."

At least that got him talking. "Hey," he said, softening somewhat. "As much as my customers try to fuck me, it's a wonder I'm *not* knocked up."

So then I said, and then he said, and then we said, and said and said and said—past the hospital and on to the city limits. But no matter what the words, I couldn't pry our destination out of Ejo. That didn't happen until we turned onto a county road and motored on about two more miles. Here's where Ejo slowed down, and the reason for him taking his foot off the gas was over to the right.

Over to the right was this very large and active wrecking yard. Surrounding the place were farms and fields, and that particular quiet country presence in contrast with the purpose and activity of the wrecking yard made the yard seem as some sort of separate domain. At first glance, the sight of it made you think of it as a "Realm of Wreckage," but then when you looked again, the apparent sovereignty of the scene and the way its purpose took hold of your senses made you update your thinking to the point where the place seemed much better described simply and wholly as the "Wreckdom."

When Ejo turned down the lane leading to the gate, that made me say, "Don't tell me you're junking this truck. It still has four round tires and half a tank a gas."

He saved his answer for the scale house. The scale house was the nerve center of the Wreckdom. There was an inside scale for people bringing in the small stuff, and a big drive-on jobbie outside for loads hauled in on big trucks. The scale man tallied up the weights and noted the contents, and his secretary ran the numbers and disbursed the checks. He stood, she had a desk. Separating them from the customers was a counter. The first thing Ejo did when we walked in was take a clipboard off the counter and make a quick signing on the paper it held.

When he handed the clipboard to me, it was plain to see that what we were signing was a disclaimer. Translated from legalese, it merely stated that a signature was required as a passport to the Wreckdom, and that the signer released the Wreckdom of responsibility for any shit that might come down inside. To guys like me and Ejo, we'd take one look at such a disclaimer and then look at each other and say, "Well, duh," but at the same time we signed, we were also aware of the lurking presence of those people still in the larval stage who think that filing phony lawsuits after faking injury or deliberately getting hurt will actually grow them a pair of wings.

So much for the legal end of the sign-up sheet.

On to the practical.

Signing the sheet allowed us entry into the Wreckdom.

And why would we want to enter such a place? A better question might be "Why not?" because even though the Wreckdom was a junkyard by another name, all the so-called "junk" out there was for sale and all of it was on sale.

Ejo was proud to bring this to my attention. He walked ahead of me, making grand, sweeping gestures with his left hand and using his right hand to point out specifics. "If it's any kind of pipes you want, there they are over there. Angle irons too, and if you need I-beams, channel irons or just plain flat stock, it's all in the same pile. 18 cents a pound. Last month they jacked it up from 15 cents but that's still only a tenth of what you'd pay new."

The tour took maybe twenty minutes. We stopped by heaps of cast iron ("26 cents a pound"), checked out huge mountains of aluminum ("Need a storm door?—41 cents a pound") and walked between rows of junked cars. ("I get a lot of my parts out here.") Ejo had to talk loud while explaining the layout and quoting the prices because none of the business conducted there was done in silence. A huge crane with giant claws was lifting tangles of metal from a semi-trailer, swinging the tangle toward a similar tangle of metal lying in the yard and then letting go at the end of the swing. The clang and clatter of this operation in concert with the same kind of sound put out by the operations of an electromagnetic crane and related racket coming from cars being crushed made the loudest sound you could make with your mouth compare with a whisper.

Ejo saved the best till last. After making his way past several big bins holding engine blocks, and other, smaller bins full of transmissions and gear drives, he stopped at another bin labeled "Stainless Steel" and said, "See anything in there you can use?"

The bin was one of those extra-large dumpster types that you winch up onto a tilt-bed truck whose back tips down behind in order to pull up the load. To look over the top of the bin was to see literally tons of stainless steel in all shapes, forms, and sizes. Bars, tubes, pipes, plate and sheet metal, springs, gears, rods—all tossed in over the top, some of the pieces lying neatly in parallel but most having that "scattered look" of a trailer park hit by a tornado.

While I stood there noting the condition of some of

the stuff inside, and while I was wondering how so much of the better stock could be considered "scrap," Ejo looked at his watch and said, "It's getting late. If we don't get back to the shop before ten I'm going to miss my smoke break."

Not wanting to deprive him of one of life's Great Toxic Pleasures, I quickly pictured what the Bus needed in the line of rust replacement and picked out several sheets of thin-gauge stainless.

On the way to the scale house Ejo put me on notice. "I hope you brought along your bank book and your first-born son, because what you have in your hand there goes for the grand total of 50 cents a pound."

He was having too much fun with my overpayment for that one piece back at the shop, so I did my best to have some fun of my own. "That's why I brung you along," I said, patting his back. "To co-sign for a loan."

When we left, my mind was still there. It was on rewind, doing a recap of the tour, first signing in, and then going through the yard. Man, the stuff in that yard. Not just raw material, but whole things—whole bikes, lawn mowers, whole desks and file cabinets, and countless other good and better things having nothing missing except someone's desire to keep them. This was both a shame and an opportunity, and all I could think of was, "I have struck the mother lode. I have struck the mother lode, the father lode, and loads and loads of Possibility without a premium price."

To understand the role of Possibility, all you had to do was walk among the "junk." Say for instance you've got a Splitwindow Volkswagen Microbus with a muffler that needs replacement every two years, mainly because it lives in the bottom back of the Bus, it's made chronically thin and cheap, and suffers from a spray of salt water every time the rear tires splash through a winter puddle. Now you, tired of having replaced at least six or maybe even half a dozen mufflers already—you look down into the bin marked "Stainless Steel," and what do you know. Along with everything else unimaginable, you see tubes, and some of these tubes equal the size of the VW muffler body, so you pull one of them out, note that this is a thing built to last, and carry it up to the scale. And there the tube, plus a few other pieces you pulled out to serve as fittings and baffles, comes to something like six dollars and fifty cents, which at that time was one-fifth the price of a standard ferrous metal muffler. I say, "at that time" because that was three dogs ago, and since then the value of the dollars that bought the stainless material have rusted away to a fraction of their former worth, but the muffler they bought still shines in both function and appearance, just as it did on the day it was bolted on.

    Over time, almost everything in or on the Bus had its origin in the Wreckdom. Here's where some will say "including the driver," but snide remarks aside, the list starts with the bike rack in the front and ends with the bumper in the back, and in the middle there sits a wood-burning stove side-by-side with a double-burner propane hotplate put together with parts from a gas stove that somebody drove into the Wreckdom and pushed off the back of a truck. That same stove provided the raw materials for a deep fryer, because there's nothing like the taste and smell of a heap of sizzling fillets piled up on a big plate. And then too, most of those fillets—still attached to the fish—were pulled up fresh through an opening in the floor, something easily done when parked out on lake ice, since the opening, an eight-inch square lined with sheet stainless, was specifically designed for drilling and chilling. Lift the cap off the opening—also pulled out of the Wreckdom—drill you a hole, and once that's done is you downloads da bait and you uploads da fishes, all the while sitting in a lawn chair converted to a "webbed seat,"—that too fresh from the Wreckdom. I could mention that the roof rack was another thing fashioned from raw material rescued from the Wreckdom, and I could also claim that the hinges and the hardware for the pop-up windows were scooped up from the same place, along with the food racks, the cast iron cookware, and even the frame for the bunk. I could mention all that and then add that the rewind rope-start had its origin there too, but I won't, and neither will I take time to bring up the fact that the vent through the roof and the catches for the canoe tie-downs came from things thrown away. None of that will be mentioned, and what will also not be mentioned is the source of the jack and the jack stands, because it's time to get on to more important things, like beer breaks, walks in the woods, some serious nose-picking on the value of both unity and division, and finally, sympathy and regret for the sickening state of the Great Body of Humanity.

    Along with raw material and Possibility, the Wreckdom was also chock full of Suggestions. Parts would pair with one another. A gear might say, "Remember that shaft you saw on the other side of the pile? Wouldn't it be great if I fit on that shaft and together we made the basis for a contraption that had the capacity to kick-start the motor on your Bus?" And sure enough, with that suggestion the shaft was bought and so was the gear, and those two items along with one more gear and a few bearings and braces became a bolt-on alternative to starting the Bus with the key. Coincidentally, this was soon after the rewind rope-start gave off its last gasp. After more than a dozen years of faithful service, instead of bringing the motor to life with one pull, the rope-start responded with a death rattle. An autopsy revealed parts in the interior to be broken beyond repair, so I took the whole assembly back to the Wreckdom, said a few words in praise of the rescues it gave and the memories it made, and then went back home and did what the gear suggested. Soon I had a Splitwindow Volkswagen Microbus whose motor started by putting a foot on a pedal and giving a casual downward push.

Something was wrong. I mean it was too easy. Starting the motor this way was way too easy to believe, so I kept turning it off and restarting it to give all my doubts a chance to prove themselves. But when more than a dozen easy starts left all doubts in the dust, I had to concede that yes, the kick-start actually worked and it worked better than the rope-start, simply for the fact that it functioned by a push of the foot. Instead of using your arms to pull up and out on a rope—arms more suited to lifting coffee cups and bottles of beer—now you're pushing with a foot. You're pushing *down* with a foot. Here you've got weight, strength, and gravity on your side, and those three plusses give you twice the torque with half the effort. Just another example to show that you can't stop progress.

Suggestions didn't stop there. A set of steel rods implied they could better serve in the Bus as clothes dryers and boot racks rather than something to trip over in the Wreckdom. Flexible metal hoses hinted they could help cool the hot spots on the motor by rerouting the flow of air, and a throttle from a tractor and bearings from a file drawer put me wise to the possibility that, together, the three of us could build a better thermostat for the motor than the one put on back in Germany. The list goes on and on—on to the extensible smokestack, the rolling rooftop rod to help load the canoe, then to the holder for the grilling stick, and finally to the oarlocks. These were not ordinary oarlocks. These were not those clunky, pin-and-socket mothersluggers that make more noise than motion. No, these were suggested by a snowmobile, of all things, and of all things relegated to the Wreckdom, snowmobiles reign supreme in serving up suggestions. This time it happened while I was checking out the front end, specifically the steering mechanism. And there, up front and out in front, a tie rod and a ball joint both said, "Never mind us how we are. Think as we could be." And sure enough, acting on those words produced a set of oarlocks that moved my canoe with a kind of quiet grace and elegance that was not fully appreciated until a crisp morning in October when the silent operation of the oars allowed me to inadvertently sneak up on a naked woman.

(More about that later.)

So the Wreckdom wasn't just some sort of metal mortuary. Ejo called it the Paradise of Parts, and I often thought of it as the Location of Innovation. Mostly though, it registered with me as the Wreckdom, since its fenced-in feel of heavy industry made it seem like a loud and sovereign domain existing starkly apart from the tranquility of the surrounding farms and fields, and since its basic purpose, after all, was to be the depository of what most members of modern society would casually dismiss as "a bunch of shit."

Ejo lived his hundred years in a hurry. When he wasn't filling his days, he crammed them. Every time he got a speeding ticket or a ticket for going too fast, he made a deal for a different set of wheels. "Cops know you by what you drive," he would say. "Get a ticket, get a different set of wheels."

The last time I saw him he was driving a white van—fast. He came up from behind, passed, and as he passed he flashed a big grin and followed it with a speedy wave. That was one day. The next day he was dead.

Before the funeral flowers had a chance to wilt, I backed over my bike. That is, I backed over my bike again. It was part of an ongoing habit started years earlier and went like this: I'd park my bike in the driveway behind the Bus. Later—forgetting the bike was there—I'd hop in the Bus and back up until I heard the sound of crunching, which was itself followed by the sound of profanity. You'd think a guy would learn. After how many times? This time it was the back wheel. The back wheel was bent in a way that would never let it pass through the rear fork again. Here we'd be talking high-dollar replacement if not for what Ejo called the Paradise of Parts. Out there would be many replacement assemblies to choose from. There always was. I'd pick out the best, and that premium part would retail for 18 cents on the pound. No wonder I never learn. Some people only learn when they pay.

I checked my wallet, and when I saw it had the few dollars necessary to cover the cost, I backed the Bus out of the driveway and puttered down the road to You Know Where. Soon it was in view and seemed no different than before. It rose above the fence-line like an eyesore or an opportunity, depending on your own horizon. It was either a bad neighbor or a great natural resource, a ten-acre wrecktangle of junk and noise or a gold mine of parts and possibilities.

When I turned down the lane, I wondered what I'd

find besides the wheel for the bike. Probably more than I could carry. Already I could see myself struggling back from the yard with armloads of "junk," some of it even having precedence over the bike wheel itself. It happened all the time—go for something specific, something needed today—and come away staggering with a load of equally staggering possibilities for tomorrow. It was exciting, was what it was, partly because of the time gone by since my last visit, but mostly because of the country I was in. Was this a great country or what? I got all gushy-feely with the idea that I was living in a country structured to excite a person's sense of possibility, knowing that filling people with an excitement of the possible was a critical component in the makeup of the country itself, almost as if each citizen was granted status as a stem cell, a cell born free to differentiate—differentiate into art, science, medicine, crime, or you-name-it, and if you wanted to name it, then maybe the whole bunch of us pooling our talents and acting in concert could rightly be called the Great Body of Humanity, because now, with our collective effort, look what happened. Active cooperation in the pursuit of possibility has put people in control of a creature greater than the sum of themselves. This Great Body has more muscle, sinew, brain, and bone than the sum of its own cells because the cells pooling their talent and working in concert can make their collectively created body give each of them much more than they could gain by only working alone. No doubt about it. It was exciting, and as long as the excitement centered on a sense of possibility, it was plenty safe to say that you can't stop progress.

    All that lofty thinking popped like a soap bubble bumped by a cigarette when I walked into the scale house. Because the signup sheet was gone. In place of the familiar clipboard holding a sheet of willing signatures was a notice taped to the counter. It read: "Effective immediately, there will be no more sales from the yard."

I tried not to sound stunned when I asked the scale man the reason for the change. "Well," he said, with that special tone of resignation you hear from the helpless, "We talked to our insurance company and they said allowing sales from the yard was getting to be too dangerous."

"Did something happen out there?"

"Well, no, but our insurance company said the risk was there."

"But that's what insurance companies are for. To cover the risk. Isn't that their job? I mean, why pay them if they're not doing their job?"

When the scale man gave his answer in the form of a shrug and a slow shake of the head, it was plain to see that what we had here was a funeral with no hope for a resurrection.

Still, with no hope for resurrection, correction, or insurrection—even at that there was still hope. That is, there was still hope for somebody to make a blathering fool of himself. Also this was an excellent opportunity for the same guy to waste time asking why the ones who amass the most amount of dollars seem to have the least amount of sense, so, not wanting to knock opportunity when opportunity knocks, I launched into a series of loopy questions and rambling diatribes that left no doubt on the part of the scale man that the person he had in front of him was truly crazy or was long overdue for a double shot of good whiskey followed by a quick beer. I mean who else but a genuine head case two pills short of a prescription would go on and *on* about some redneck renaissance man—some guy he called "Ejo," and how the next generation of Ejos would remain bullied little fat kids—all because Big Money had the power to close the *rectum?* Weird.

When I left, I left with the idea that I wasn't done yet. That is, I left with the idea that I needed to make an even bigger fool of myself by contacting the company hired to insure the Wreckdom. Only instead of firing off a letter of indignation that would go directly from the mailbox to the shredder, I decided to go directly to the CEO himself. Of course you just don't do that sort of thing. At least you don't do it in the standard way. So instead of seeking out his corporate office and failing to penetrate his iron ring of secretaries and security, I had him come to me. It was better this way. This way I would have a better chance to confront him about closing the Location of Innovation. Maybe I could even shame him into opening it back up because there we would be, both in the Bus, him riding shotgun, two guys tooling along in a virtual showcase of possibility, and with no danger of the obvious being made purposely vague by clarifications from legal advisors.

But when he accepted my invitation and sat down, I got more than a double surprise. Instead of some big-bellied old flabbyface in a pricey suit, bald and jowly and reeking of cigar smoke, there he sat, young and clean, smiling and relaxed, and with shirtsleeves rolled up and tie hanging loose. He was not looking one bit like some senior executive about to be shamed into changing his cancellation policies. No way, so instead of opening the conversation with an accusation, I simply said, "Nice day."

"Beautiful," he said, and went on with, "And you know, these windows you have here kind of add to it. Look. I can slide this side one back and rest my elbow, and this wing window, by swinging it out, it's like turning on a fan."

"Early German air conditioning."

"Seriously. You don't see that anymore. Cars built now are designed to be closed up."

"It's not just cars," I said, going from there. "It's a lot of stuff. Junkyards, for instance." Here I went into a repeat of what I said to the scale man—without the rambling vehemence—adding that by closing a potent place of possibility this was cause for all the upcoming Edisons and anyone else wanting to grow wings to remain cocooned, held in suffocating bubblewrap as not much more than pale grubs, stunted by a drug sold as "pseudo-safety," and sold by insurance companies like his not having the balls to do their job.

"But we are doing our job."

"How can you say that with a straight face?"

"Our job is to make money."

"Well, duh, but to do your job you have to insure against risk."

"No, no. You don't get it. Our job is to *make* money, not earn it. There's a difference."

When I just looked at him, he went on with, "Earning money is obsolete. It's too much work. So as insurers, when somebody makes a claim, now we just cancel them. Back in the dumb old days we paid claims, but now that we're smarter we just hit the cancel button. Pretty soon the word gets around. Make a claim, get dropped. People run in fear at the thought of being dropped, so they pay their own claims. They keep paying on the policy plus they pay their own claims. Beautiful."

"You mean pretty. As in pretty much like extortion."

"There's more. Since we no longer have to pay claims, we take all the money from the policyholders and throw it at high-risk, high-yield scams. If the money sticks, we make money, but if it doesn't stick, then we get all of our money back to throw at other scams."

"I think you lost me on that last part."

"Oops. That's because I forgot to tell you that the scams are backed by the government. So if the bottom falls out, we still get our money."

"From the government. You get your money from the government. But think about it. Where does the government get its money?"

"Of course. From the same people who buy our intimidating policies, only now we're fleecing more of them, and some of them twice. But who cares? It's business. Business is all about moving money. It used to be about moving goods and services to make money but now it's only about moving money, and the more money you can move into your own pocket the more successful the business."

Somehow this was not going my way. I was supposed to be asking burning questions and he was supposed to be sweating from all the heat, but as it turned out, I came across as a sputtering ball of confusion while he sat there smiling, relaxed, and proud. *Proud.*

A phone call rescued me from further embarrassment. One of his colleagues called from Washington, thanking him for coming through with a large order of lubrication needed for the greasing of vital political machinery, and with that particular news he excused himself and quickly left, explaining that the phone call had him so dizzily excited that he absolutely positively had to return to his corporate office, uncork a bottle of champagne, and jerk off.

"What?"

"You wouldn't understand," he said, without looking back.

I was glad he left when he did—relieved too, because there was no telling what might have happened next. I was afraid he would explode. I mean you had to be there. Here I thought he'd be ashamed when he described his so-called business plan, or at least apologetic, but no, all he showed was pride, and lots of it. They say pride goeth before destruction, and the way the prideful get themselves destructed is probably the same way a one-pound bag gets destructed by filling up with two pounds of shit. It explodes. No way would I want that to happen in the Bus. It's bad enough when somebody cranks out a big corned-beef-and-cabbage beer fart and you have to break the windshield for relief, so I felt no sense of sorrow when he left.

Any sense of sorrow was saved for the Great Body of Humanity. Now there was a creature in peril. Driving along, and once again driving alone, it was hard to shake the feeling that part of this Great Body was dead. A vital part had bit the dust, and in its place was a tumor. Brain cells and nerve cells that were basically the guiding light of the Great Body of Humanity had been converted and perverted into highly destructive political, financial, and legal cells. Essentially they were tumors, these groups of cells, and as such their only function was self-enrichment, using all the supporting cells of muscle, heart, blood and bone to either contribute to the cancerous effort or get out of the way. Here you want to ask the malignant mass, "What does a cannibal gain by eating itself?" but as long as the tumor is blindly growing, its proud and happy answer will always be, "You can't stop progress."

Four Dogs and Two Wars Later

"Never judge a hippie until you walk a mile in his bare feet."

(Taken from the last chapter of the
Redneck Bible, the Book of Reservations.)

"How many miles per gallon does it get?" Drivers of early VWs heard that question all the time. Because back when the Wolfsburg factory was cranking out classic Volkswagens like popcorn popping out of an open pan, good gas mileage was a major selling point. Almost every ad had an economy connection. Even though this was a time when gas was cheap, and even though this was a time when most of what came off the assembly lines in Detroit had a focus on hyper-muscularity or morbid obesity—even at that, or maybe because of that—there were plenty of people who were plain up-to-here with automotive overkill. They were sick of Goliath and dreamed of a David.

Someone in Germany was listening.

Because pretty soon it's like the dinosaurs being challenged by the mammals. Only this time it's the dinosaurs being challenged by a Bug. A whole new species of automobile comes into being. And not in line with the theory of evolution but with a practicality that countered the monstrosities that evolution had evoluted.

Kinda makes you wonder about the rest of it.

Anyway, here comes the Bug, followed by the Bus,

and now we're talking about the greatest automotive success story in the history of the world. Earth-friendly cars as opposed to those more suited to Uranus. Good on gas, fun, and funny. And once again, good on gas. It had that reputation. It was a well-deserved reputation, and it was common for drivers of early VWs to compare notes on gas mileage. "Oh?" one driver might say to another. "Yours gets 31 mpg? Well, believe it or not, mine gets 80,000 mpg,"—quickly adding that mpg stands for meters per gallon. So, no matter how you measured it, both the Beetle and the Bus had a reputation for being good on gas.

However, there was another reputation. It was a reputation starting in the sixties, and it was exclusive to the Bus.

"That there one a them hippie vans?" Back in the day—back when the Vietnam War was raging in Asian jungles and on American streets, and back when violent members of the majority wanted peaceful members of the minority neatly lined up into neeg rows, "That there one a them hippie vans?" came across as less of a question and more of a threat. The Status Quo had no use for the Let Us Go, and to those who considered themselves to be "in the trenches" and "holding the line," "Them hippie vans" were the rolling equivalent of a social disease. Didn't matter if you were only a simple beer-drinking former Seabee and were only on the road for the fishing and the fresh air. What mattered was, if you were on the road in a Splitwindow Volkswagen Microbus, get ready for trouble whenever you heard "That there one a them hippie vans?"

But way later, four dogs and two wars later, still beer-drinking and former military, and still rolling over the road in the same Bus and for the same reasons, I'm filling the tank at a rural gas station in the Deep South, and here comes that same question, spoken in a drawl and coming from behind.

Reflexively I thought, "Here we go again," but when I turned around and saw a man approaching with a fake swagger and a big smile, well, all I could do was put on a big smile of my own.

"Betcha never heard that before," he said, no doubt in response to my sharp turnaround.

Recovering, I said, "No, and another thing I never heard was You ain't from around here, are you."

That made us both laugh because "You ain't from around here" was another one of those loaded catchphrases. Like "That there one a them hippie vans?" "You ain't from around here" was to let you know that both you and that hippie van better *not* be around here the next time I blink. Them was the days, so to continue with the memory I said, "Know how many hippies you can get in a VW Bus?"

"At least one more and a dog."

With that, any ice that remained was broken and the pieces had melted into a common flow. We were about the same age, and apparently had gone through the same rage. His name was Lee, or so I thought, since he was wearing a dark blue mechanic's coverall with "Lee" stitched in white letters over a chest pocket.

"No, I'm Bobber," he said. "This here's Lee's soot suit. I got a big job under a dirty truck back there in the engine bay which don't make no sense stink'n up my clean one when Lee's got one hang'n that hardly can't hold no more dirt."

Lee was his brother. Nine months apart. "Fast

work," said Bobber. "Daddy never wasted no time in neither make'n us kids or slap'n us into line. I cried when he died. Lee did too, and said half his tears were sorrow and the rest were joy."

Bobber and Lee were in business together. "We bought this place when it was no more'n a rusty pump and a cluster a beer signs. An' look at 'er now," he said with a grand sweep of his arm. "Two engine bays and one big enough to hold a semi. New pumps with three grades a gas and low sulfur diesel. Kerosene's over there. Smokes and beer, we got 'em. You want ice with that beer, a buck two cents a bag, pure and clear."

"Beer?"

"Course beer. Good beer, and lots of it."

"We're not in a dry county?"

"You want a dry county you gotta cross the line and take the first road on the right. We got more sense than that here. What kinda beer you look'n at?"

"Bad beer."

When he just looked at me like someone waiting for a punch line, I did my best to explain. "I've been all over this country and north of the border trying to find a bad beer. Up, down, and in between. So far they've all been good, better, or best, but you never know. Maybe one of these days I'll get lucky."

"Sounds like my kinda quest. I'd jump in and go with you right now but my life's all here. Always has been and always will be. Now I got this place—me and Lee—and I'm good as glued to the ground. Makes me happy. Be happier yet I could call it Robert E. and Lee's Service. Think about that. You know history and you see the link, but Lee says no. Older brother you know. Shoulda not said noth'n and let him thought of it first."

"I used to know a set of twins like that. One's older by two minutes. Guess which one has to be the boss."

"Yeah, that's Lee alright. He's the family pissant, but he's got good sense otherwise. Weren't for him I'd still be put'n in my forty over at the feed store and the only way you'd be get'n gas in these parts is if you was eat'n beans."

"You did good."

"Thanks to these," he said, with a nod to the Bus. "What I'm say'n is, me and Lee made a young fortune fix'n up and sell'n these here hippie vans. Musta been a hundred went through here, if one. Course you know why. Them hippies, they can't fix noth'n. That's the differnce between hippies and long-haired sonzabitches. Long-haired sonzabitches breaks down they got a full set a tools and hands that know how to use 'em. They come in here and buy parts. Hippies come in here and want help. They got no money but they want help. Long story short, we end up with the van. Hey, it bought us this place and twenty acres behind. Figure one hundred times a thousand and that's about how many dollars me and Lee made fix'n up and sell'n hippie vans. This a '62?"

"Mostly. It's got some newer parts and some older ones. The driver's a '44."

"Thought so. 'bout the Bus I mean. '62's the year they quit them banana-tit turn signals and put on the one's like you got there. Fried eggs, they call 'em. I'm a '46 myself, but nobody believes me with my black hair and all. You 'bout ready to sell this yet?"

"Close. They say I way overpaid for this buggy, so it

still owes me some miles."

"They go forever you treat 'em right. Betcha this one's got some miles."

"More than zero and less than a brazillion."

"Good answer. Ain' supposed to know how many miles. It's all about the road, they say, and what the road don't give you the miles can't."

"Yeah. Yeah," I said. "I never thought of it that way but I think you're right. Mind if I use that later on?"

"Ain't mine to claim. Like to claim this van though. You don't see them around anymore. Used to be they was all over the place. Mind if I look inside?"

"Watch out for the dog," I said, meaning Scotty lying on his side on the bunk, one eye open and tail flopping up and down in half a wag.

"Yessir, he's a bad one." Bobber said, reaching in and scratching Scotty's ears. "He gets any badder he'll be bite'n the balls off frogs and crickets."

While Bobber looked in on the Bus, I finished filling the tank. I brought it up to where the sound of the gas going in began to rise in pitch. That meant the tank was near the top and gas was about to enter the filler pipe. No point in topping off the pipe and having gas push past the vent in the cap the next time the Bus took a hard left or failed to dodge a bump, so I stopped at the sound of the tell-tale rise in pitch.

When I hung up the nozzle, Bobber was still looking in on the Bus. "Too bad Lee ain' here. Me and Lee seen a lotta hippie vans but this is someth'n else. He won't believe me you got a whole house in here—'specially if I told it—so you hold tight while I go get me a camera."

Bobber took more than a few pictures, and wanted Scotty in almost every frame. "Put him in the driver's seat," he said. "That's right. Can you get his feet on the wheel? Now you and him, sit'n down here on the side." And finally, "Here, take one a me an' him. Stand back a piece. Make sure you get the pumps in there."

When all the pictures were snapped, and when the camera was back in Bobber's hands, he put on a face of false worry and said, "Lee's not gonna like this. Lee, he's gonna take one look at these pitchers and say Whyja let that one get away? So I gotta make you a offer. I know as a positive fact I can get a hunner dollars for this van so how's about I offer you 99?"

"Man, you sure know how to tempt a guy," I said. "But we've been friends for too long." I looked at my watch and went on with, "We've been friends ten minutes already. Suppose I sold you this old can of sauerkraut for 99 dollars and all you could get back was 50 bucks? That's no way to treat a friend."

"Okay then, 49."

"Do I get to keep the dog?"

Before he answered, Bobber took a long time to look in on Scotty. Scotty was back on the bunk, once again lying on his side and flopping his tail up and down in half a wag.

"It's the heat," I said. "It makes him lazy."

"Had a dog like this once," Bobber said, reaching in to scratch Scotty's ears. "Loved that dog like my own baby. Fella says to me, You wanna sell that dog? An' I says Take a thousand bucks and put a ten in front a that an' you still ain't got the down payment. Yeah. Loved that dog. A dog's a man's best friend, they say. They got that right. What's wrong though, is when you hear tell only a man got a soul. No sir, not true. You wanna count soul on the same page you count good, then a dog got more soul in her short life than a man got in all a his." Nodding at his own words, he added thoughtfully, "No, you best keep him. He's got that look. He's got the look of a good dog in a good home so you best keep this van too." At that, Bobber straightened up and was instantly back to his original self. "You want beer with that gas? How 'bout ice? A buck two cents a bag, pure and clear."

When I left, I left with a full tank of gas, a six- pack of good beer, and a bag of ice—"pure and clear." The gas, the beer, and the ice were gone in a matter of days, but what remained was the memory of a man who knew the meaning of worth, whether in dogs, gas, or beer—or even something as simple as one a them there hippie vans.

Do What Must Be Dung

"When the choice is either taking a shit or leaving it, always leave it."

(Olde scatological vulgarism)

It happens to everyone. There are no exceptions. Kings and queens, the young and old, rich and poor, the pretty and the pretty ugly, black, white, red, yellow, brown, and everyone in between, happy, sad, or serene. Yup, no matter where you live or who you are, periodically, peristaltically, it happens. Like truth, juice, and the caboose on Granny's goose, it always comes out in the end.

This time it was in Quebec, on the way to Montreal. It was late morning in early spring, and traffic was almost nonexistent. The road was clean and dry, flat, two-lane and rural, and had snowbanks on both sides. The snowbanks, made of old and new snow, were impressively high and cliff-like, and blocked the view of the woods beyond.

Mary Teal was with me. I was driving, she was wingman. It was our third day on the road. The length of the drive, the drone of the motor, and a lack of changing scenery had put us both in silent mode, content to cruise along without a word, comfortable with the thought that some parts of a long drive are only about the miles.

And then my daughter, the beautiful and gifted Mary

Teal, tall, blond and willowy, articulate and intelligent, linguistically advanced and twice nominated for the Nobel Peace Prize in Elegance of Expression, broke the silence with "Dad, I hafta take a shit."

And then me, no stranger to such blunt and graphic utterings made by both Mary Teal and her brother Beano—I said, "Yeah. There's a town up here about twenty miles. You can go while I gas up."

"No, Dad. You don't understand. I hafta go right *now*."

"Do you think you can wait till I pull off to the side of the road?"

"It's not funny," she said, squirming.

While coasting to a stop I thought, Ordinarily a situation like this is, shall we say, "rectified" by grabbing up a roll of pooper paper and tiptoeing off into the woods.

However, with the height of the snowbanks on a level with the top of the Bus, and with the depth of the snow in the woods up past the door handles, you needed to be a snowshoe rabbit or a moose on stilts to even *walk* in the woods, much less squat.

Not to worry.

No, no need to ruin your day or your pants with a botched notch in the back of your crotch because what was tucked under the front seat of the Bus was the prototype of the Fold-up, Zip-Sack, Crap Trap.

Never mind the name. Think "rescue," since just as fast as a paramedic can pop out of the back of an ambulance, so could the Zip-Sack Crap Trap pop out to rescue the back end of the beautiful Mary Teal.

Thank you Lord it was already loaded. By "loaded" I mean the Crap Trap already had the bag in place. The bag was a simple zip-type plastic jobbie with the open top folded outward over an oval wire fastened under the seat. The seat was supported by a set of snap-out legs, so it was just a matter of popping them into position to accommodate my now desperate daughter.

By then I was out of the Bus. I was out of the Bus and scuttling down the shoulder of the road hoping that someday Mary Teal and I would meet in heaven just in case all these preparations were a little too late and the heat generated from her urgent condition was about to throw her into a state of Spontaneous Human Explosion.

Finally far enough down the track to be safe from a fatal flame-out or a crippling blast, I slowed to a walk and further let off on the walk until I was moving along in a reflective mosey. It was one of those easy-going amblings that had more to do with memory than motion, and the memories that prevailed all fell under the heading of Narrow Excapes. Not "escapes," because at those very critical moments there was no time to quibble about proper grammatical usage. Plainly and simply they were Narrow Excapes, specifically the kind narrowly avoiding an involuntary voiding of the contents of the colon when you's on the road a roll'n. I thought back to Arkansas, and clicked on to the time that featured The Local Food, the Turnout, and the Cop. The bad part was the food, the good part was the turnout, and the best part was the cop, especially since the cop was a young cop, a cocky cop, and a cop unaware that his role was less about police work and more about being a star player in a short flick that could be titled "Close Encounters of the Turd Kind."

Not that I planned it that way. It just happened, starting with gastric rumblings. This after a greasy meal at a rural diner. I'm tooling along on a state road after filling up on a big plate of rancid pork and spiced gravy, when all at once there was this sound of thunder. It was internal thunder, and by the sound of it I knew I was only moments away from feeling the full fury of a category 5 shit storm.

This was no time to say, "so what." Immediately I slowed down and looked for a place to stop.

But this was Arkansas, a state where the roads are not bounded by wide gravel shoulders. No, they're bounded by narrow strips of poison ivy sloping sharply down into a ditch—which instantly brought to my mind the title of that old sixties song "Why don't we do it in the road."

Just before a profound sense of urgency gave credence to such an option, I crested a rise and there down in the dip ahead was a gravel turnout.

Quickly pulling into the turnout, I just as quickly scrambled to the middle of the Bus, pushed my pants down to my ankles, and sat on the floor.

But not before flipping back the cover over the ice fishing hole. Yes, the ice fishing hole. So many things besides hard-water fishing can be done through that hole. Now I was doing one of them, and what resulted on the ground below was an expanding puddle of green and yellow matter hissing like cats and snakes and bubbling like champagne made not from vintage grapes but from a bad batch of dingleberries.

We shall not mention the odor.

Not wanting to spend any more time than necessary in such a suffocating presence, I quickly cleaned up ye olde bottom end and threw the pooper paper in the stove. Then I flipped the cover back over the hole and staggered back to the driver's seat.

Talk about relief. For a minute I just sat there, semi-exhausted, feeling like a character in a bad science fiction movie, someone who had given birth to an evil demon and was greatly relieved that the demon had been born dead. (Or at least smelled that way.)

Soon I was eager to move on, but didn't, fearing that the evil demon might have an evil twin. Sometimes that happens. You know the drill. You think the threat is over when all at once you get this jolt. Here we go again. The evil demon has a sibling on the way, or if not that, then what's about to happen is an outpouring of afterbirth from the first one.

Almost as if on cue, here comes a cop. I'm parked facing west, and a car with the insignia of the Arkansas State Patrol cruises into view heading east. Even before he slowed down I knew he was going to stop. A little bird told me. He slowed down and pulled to my side of the road so when he stopped we were window-to-window and face-to-face.

Since we both had our windows open, the face he saw was of someone silently giving thanks to fate for not scheduling our meeting any sooner, and the face I saw was a double view of my own, because the cop was wearing that brand of reflective sunglasses with the silvery lenses that function as one-way mirrors—mirrors convex in shape so when you see your own distorted image looking back at you, your nose is as big as your entire head and your lips have that inflated look of a tractor tire folded in half.

Just when I was thinking of the difficulty such a face would have in the quest for a Friday night hookup—

except for maybe them lips—the cop said, "I'm sorry, sir, but I need this space to monitor traffic, so I'm going to have to ask you to move."

Move.

Resisting the urge to say, "But I already moved, big time," I simply gave him my best smile and said, "Oh," and with that I flipped the key, put the Bus in gear, and gently drove away.

Long after I left the scene my mind was still there. Think about it. You put down a puddle of poop in a place where a cop puts up a speed trap. He's a young cop, a cocky cop, and a cop wearing one-way shades. To top it off, he's chewing gum. He's not so cocky as to crack the gum while he's chewing it, but he does have the gum in his mouth and the way he keeps working it gives you the impression that he thinks he's hot stuff and gives you the added impression that he thinks you too should think he's hot stuff.

Now put that next to the fact that Officer Hotstuff has parked his golden chariot over a place where a common peasant has just let loose with a malodorous discharge having the greasy capacity to derail a train. I watched him pull in. As I drove off, I watched through the rear-view mirror and saw him move in and park. He parked directly over the magic spot. Whew, I thought, wiping my forehead, relieved that he didn't park *next* to it—like with it right below his open window so when his nose told his eyes to look down, there the puddle would be, grinning back at him. No way would I want that—at least not right away because cops are trained to prioritize and even a rookie would realize that chasing down the creator of a toxic waste dump has priority over maintaining a simple speed trap.

Later though, the thought of a cocky cop parked next

to the noxious mass came across as just fine. Peachy, really, because later (after I was safely across the border and hiding in the back streets of Guatemala), the thought of Officer Hotstuff parked next to the recent anal discharge was no longer mortifying but exciting. I pictured speeders coming over the rise and into his trap and after each time of chasing them down and writing a ticket, there he was, returning to the area of infamy. Even though I was only fantasizing, I knew it was not beyond the reach of fate that during one of his returns he would run through the pile with a tire and foul the tread—or better yet—he'd step in it. I pictured him parked next to the unholy mess, and stepping out to stretch or take a leak. Of course when he'd put a foot outside the car the sole of his shoe would come in contact with something other than "the ground."

Before my inner imbecile could take it from there and picture him not only stepping in it but also slipping and falling down, a voice from the past—an admonishing voice—came through with the words, "Remember that similar situation in Illinois? When you were caught with your pants down? You weren't so smart then, were you?"

Well, no, but it wasn't like there were any cops involved or that I was trying to pull off some sort of excremental prank. I was only trying to do what must be dung, but this time, instead of soupy poopie, the problem was centered around the eviction of its stubborn cousin, stony baloney.

The problem started back in 1959. Or at least it seemed that way. When I counted backwards on my fingers it seemed like the last time I launched a submarine was somewhere back in the early part of 1959. I knew that couldn't possibly be right, since the world did not end as experts had predicted it would at the turn of the century. Still though, 1959 *felt* that it might be right in that it felt like whatever was in there had more in common with geology than with the tenacious tenure of a tardy turd.

Fearing that waiting any longer might put up a permanent roadblock on the exit lane of the Hershey Highway, I started looking for a private place to park the Bus.

You look around in Illinois and most of what you see is one gigantic cornfield. People mainly think "Chicago" when they think of Illinois, but Chicago is not much more than a zit on the face of humanity when compared to the area of Illinois planted in corn.

Which makes it the ideal place to stage a performance where you really don't want the attention of an audience.

However, with the contents of my colon packed down into a condition that the Chinese call "hung chow," this was not going to be one of those breezy deeds where you tiptoe off into the tall stalks of corn, do a quick curtsy, and come out smiling. No, this was serious. Past experience told me that a stubborn case of costive coagulation involves four time-consuming stages: Labor, Dilation, Evacuation, and Recovery, so in effect this process combines the science of obstetrics with the procedures necessary when dealing with the aftermath of a natural disaster.

Turning onto a rural road with no centerline and no traffic, I found an open patch in the endless corn that appeared to have once been someone's yard. I stopped, and sure enough, at one time it had been someone's yard. A house was on the property. It was a small house, old and paintless, and was made almost invisible by a thick growth of boxelder trees arching up from all sides of the foundation.

My second thought was that this house must have been "a starter home" for an area grain farmer. Back in the day, back when it was common for congress to violate treaties to accommodate the progress promised by the homestead act, farmers started out small. Small was big back then, and to plant and pick 160 acres with a team of mules took all of one's time and most of one's effort. But when mules were made obsolete by machinery—inexhaustible machinery that could easily out-perform even Olympic Gold-Medalist Mules on meth, here's where most farmers took advantage of Manifest Destiny and tapped into the American Dream by expanding their holdings. They kept adding to the acreage, and kept buying bigger and bigger machines until the size of those creatures "needed the original 160 acres *just to turn the damn things around,*" as Uncle Gary once said. Naturally, in keeping with the size thing, the little old farmhouse in the vale is abandoned in favor of a big new house on the hill.

Such was my second thought.

My first thought was: what a perfect place to ditch the adobe doo-dah.

So I parked in the yard, made my way to the middle of the bus, and popped the lid off the multi-purpose hole in the floor. Why not? There was no outhouse on the property, and there was no way I could squat in the surrounding cornfield for the length of time it would take to rid my belly of the bedrock imbedded in my bottom end.

Soon I'm sitting down over the hole, with the aperture at the lower end of my alimentary canal in a state of impending dilation, periodically priming itself by giving off with short snorts and gruff puffs of gas. Correspondingly, there was movement inside, as if whatever was in there heard suspicious noises in the basement and was cautiously creeping down the stairs to investigate.

Slowly, incrementally and excrementally, a former state of emergency gave way to a current state of emergence. I felt like a landlord giving the bum's rush to a deadbeat tenant. "There," I said, "You got your head out the door. Now, don't look back. Just keep on going."

That was the good news.

As for the bad news, here comes a truck. This is where I asked God to please have the driver keep his foot on the accelerator. Typically though, God answered my prayer by having the driver put his foot on the brake.

From my position on the floor I could see the truck pull to a stop. The side windows on the Bus were low enough to let me look back at the road so when I rotated my noggin and craned my neck I could not only see the truck come to a stop but I could also see what kind of truck it was. It was a big diesel pickup, mud-splattered, and with a 55-gallon drum standing upright in the bed. The drum had a pump attached to the top—one of those hand-crank liquid transfer jobbies—so by all outward appearances it was very safe to assume that this particular truck was a farm truck, that a farmer was driving it, and now he was stopping to see why that old Split-window Volkswagen Microbus was parked on his land.

Taking it from there, it was also safe to assume that I was only moments away from suffering even more embarrassment than that time I was with my mother and she unwittingly said in the close presence of the Superintendent of Schools, "Yeah, that (name of the Superintendent of Schools) got a little better looking when he was older, but when he was young he sure was an ugly sonofabitch." I only hoped the farmer had a gun. I wanted this to be over as quickly as possible and what could be quicker than a speeding bullet.

But no, the only arms he had were the two hanging at his side, and they were the kind of arms that looked like they were born strong and got added to by his line of work. Maybe he was forty. I was sitting with my pants down over a hole in the floor and looking up at him through an open side window, and there he was, standing outside the same window with his pants up and looking down at me. Such a situation would have been a gold, silver, and platinum opportunity for him to show outrage ("What the fuck are you *doing*"), or at least have some fun with me. ("Now I'm going to stand right here with my hands on my hips until you take this Bus, your ass, *and that goddamn turd* off my property!")

But while he looked down at me from a superior position, he did not look down at me with a superior attitude, which showed when he said without a trace of irony, "Is everything all right?"

Since I had already crossed myself and had mentally said goodbye to all my friends and family, his question "Is everything alright?" came across as confusing, so I did what confused people do when their heads are muddled and just mumbled something confusing.

Linguists could debate forever on the meaning of that mumble, but proving that farmers are smarter than linguists, the farmer simply nodded knowingly and said, "I just stopped to make sure everything was all right." There was a definite tone of gracious finality in that statement and the next thing I heard was the sound of a truck driving away.

Everything had happened too fast. Well, not everything, since I was still not done with doing what must be dung, but other than that, all the events surrounding the encounter with the farmer came at me like a flurry of punches when you're in a fist fight while drunk. Maybe it was my expectations. Sometimes expectations can throw you. You believe in an outcome, and then when the outcome contradicts your belief, you absolutely positively cannot figure out how a tiny little bit of truth can trump your deep-seated belief.

Later, while still trying to clear this up, I scheduled a consultation with the family know-it-all. Every family has a know-it-all. Usually this function is fulfilled by an in-law. Sometimes it's someone in the immediate family, but more often you marry into it. True, know-it-alls can be annoying. However, if you need an honest answer more than you need sympathetic agreement, go with the one who knows everything.

True to type, the family know-it-all was quick with an answer. When I asked why the farmer didn't make a big stink over the fact that I was trespassing *and* littering, she said, "He felt sorry for you."

"You mean he…"

"No, not that. He felt sorry for the way you look."

"I'll have to admit, sitting on the floor like that, I probably looked…"

"No. You're not listening. I said the way you look. The way you always look. Half the time you look like you crawled out from under a rock and the rest of the time you look like you should've stayed there. It doesn't help that you drive that old Bus."

"But do you think maybe…"

"No."

"that he was…"

"No.

"well, just a nice guy?"

"I said no. There are no nice guys. Believe me. I've seen them all."

There was some doubt as to whether she was speaking literally or metaphorically about having seen them all, but there was no doubt—at least among and beyond the immediate family—that she knew everything.

So, to reconcile my belief that the farmer was a nice guy with the fact that the family know-it-all said there were no nice guys, there was only one way to bring this contradiction to a logical concussion.

The farmer was dead.

Apparently, sometime after I last saw him and before consulting with the family know-it-all, the farmer, the last of the nice guys, the final male member of mankind with no measurable amount of malfeasance, bit the dust. And of course the internal antenna of the oracular family know-it-all instantly picked up on his demise, filed it, and knowingly passed it on to me. Sad, too bad, but iron clad. Wanna make three score and ten? Never question the veracity of the family know-it-all. Better to think "safety first."

Anyway, thinking it was safe to think that all this reflection had given my daughter enough time to empty the enteron and button up the bumgut, I strolled back to the Bus only to see her arms waving from the inside. She was waving me away. Evidently, even though the bus was parked, something inside was still on the move. Either that or she was finally finished and was taking time to allow all the pertinent parts to pucker back into their pre-poopie positions.

Regardless, it was plain to see that she wanted privacy while whatever was happening was happening, so I called out, "Just checking," turned away, and once again moseyed on down the shoulder of the road.

There was a breeze by then, a gentle breeze, and a breeze pure in the sense that it carried no scents. Soon, spring would change all that. Soon the snow would melt, and as it went to water there would be a release. Odors locked up by Old Man Winter would be let loose, and together they would form a familiar fragrance. But for now, only breeze, light and fresh but certainly not sterile since memories came with it, memories made possible when water turns to ice and lays down a solid surface you can drive a Bus on, drill down to water and lift out food, food that swims in water and sizzles in oil.

Look at a map of Wisconsin, Minnesota, and Michigan. Put your eye on the northern end. Is that a land dotted with lakes or is that a place of lakes dotted with land? Doesn't matter. Lotsa land, lotsa watta. Size-wise, the water ranges from a few acres all the way to lakes bigger than whole states. In this part of the country, winter turns the top part of the water in those lakes to ice, and when the ice is thick enough for walking or driving, people walk and drive. On the bigger lakes you can walk out or drive out so far as to make the land behind look like nothing more than a pale blue line, an iffy horizon that disappears in haze or snow and has you feeling as though you're standing on some sort of vast crystal prairie. It's a seasonal wilderness, is what it is, quiet and restless, beautiful and brutal, and, as a note from the dead addressed to the unwary, a patient place with a total lack of mercy.

In other words, the perfect place to set up a fish camp.

Think about it. No camping fees, no cops, and no signs warning of criminal trespass. Oh, it gets cold, but that's what makes warmth worthwhile. Wind is there too, big wind, Bus-rocking wind, and to be frightened by such a thing—well, that only means that you don't know that when the Bus hears the music of the wind its happy reaction is to dance.

And there on the shore of a big lake where the Bus likes to dance to the music of the wind, my Collie-Shepherd buddy Scotty, He-of-the-White-Feet-and-the-Tawny-Top, watched as I looped on the tire chains. I explained to him that in some circles chains are a symbol of bondage, but looped and tightened around a set of winter tires, chains represent freedom, and proved it when we both hopped in the Bus and drove several miles out on thick ice and through deep snow, with the back tires churning and turning and dynamically saying no to any static force that demanded we stop.

Since this was the crystal prairie, the rule was BYOF (Bring Your Own Firewood), and that we had in big boxes stacked in the back next to that other essential, beer. After easily finding a place with no parking meters, no fences, and no frantic consumers with runaway shopping carts smacking into the backs of your ankles, I parked and let Scotty sniff out the surrounding snowdrifts while I put on the addition. To put on the addition, what you do is you take a flat-bladed snow shovel and use it to cut blocks of snow. Snow on the crystal prairie hardly ever falls down into fluffy piles, but even if it does it's soon blown into solid domes and drifts. This makes driving tough but building easy since here you can take the blocks of snow and build yourself a Bugloo. A Bugloo is the same thing as a Bigloo which is a combination Bus-and-igloo, and the way that comes to completion is you start stacking the blocks in two parallel lines outside the cargo doors. You start stacking right tight against the side of the Bus and build outward and upward, and keep on stacking and shaping until what you have is a Split-window Volkswagen Microbus with an igloo-like addition. If the Bus had 54 square feet of living space before, now it has 70 or so, and in that extra space you can stack the firewood and bury the beer. Note to winter campers: Always bury the beer. In the night air of a northern winter even high-octane home brew will freeze and pop the caps, but bury that same beer deep in snow and the air will leave it alone.

Having one hand wrapped around a cold beer and the other resting on a fuzzy dog while lounging next to a glowing fire is the second best way to spend a winter night, but morning calls for a different brew, so, soon after the rising sun brought in the beginning of a new day, I responded to the call with a big pan of dark roast coffee which always goes good with a big plate of hash browns, toast, and eggs—eggs butter-fried and over easy—plus more coffee. Scotty never was much of a coffee freak so he washed his eggs and toast and hash browns down with water.

By then we were fishing. I had a hole drilled down through two feet of ice below the opening in the floor and was pulling up an occasional perch. When the fourth fish came up through the hole and flopped on the floor, I said to Scotty, "Guess what we're having for lunch?" I knew he knew what I meant but just in case he was having trouble with the math, I added, "At least eight fillets of deep-fried perch. You pick four and I'll have the rest."

Not all of the focus was on lunch. Some thought was saved for yesterday's meal. Yesterday's meal—the food chewed in the toothy little space behind my lips and passed through the esophagus and down into the stomach—all that was near the end of the digestive process. It had left the stomach through the duodenum and made its way through the gastro upper tunnel section (guts) where all the useful products from every mouthful were absorbed by the body to keep it happy. The key words here are "all the useful products." This implies that some of what is eaten is not useful. At least not to the body, so for that reason the alimentary canal has at its terminus an expandable opening. It's located "way back and down there," and is easily identifiable from other openings in the same vicinity by the fact of having rust-colored wrinkles around the rim. In Latin, this anatomical curiosity is known as the "yield operated unfolding repository aromatically surrendering smoothly homogenous organics largely elongate," but in common English it goes by the simple acronym, "y.o.u.r.a.s.s.h.o.l.e." Regardless of its name, it functions as the back door of the digestive apparatus and periodically allows passage of indigestibles, which by that time are collectively known in Latin (Pig Latin) as "Itshay."

What I'm getting at is, it was time. I snapped open the legs of the Fold-up Zip-sack Crap Trap and paraphrased Ecclesiastes by saying to Scotty, "To every thing there is a squeeze'n."

You'd think that would be the "end of it." Do the little awkward routine, fill out the necessary "paper work," and put it all "behind you."

But there was more. Monumentally more, because while perched on the loo, I was also holding my fishing pole. And wouldn't you know, here comes a bite. Instinctively I set the hook and began reeling in line, at the same time excitedly saying to Scotty, "Can you believe? I'm pinching a loaf and catching a fish *at the same time!*"

And then Scotty, stretched out on the bunk, partly opened one eye as if to say, "Take it outside."

No way was I going to allow his indifference dampen what I saw as an event too far above the ordinary to be dismissed as a mere coincidence. Think about it. Downloading yesterday's dinner while uploading today's lunch has a way, shall we say, of "grabbing you by the ass." It's what academics would call a "teaching moment." Oh? Yes. Because not only did it shock me out of thinking linear and propel me into thinking circular, but it also allowed deep insight into the workings of politics and organic farming, in that if it wasn't for spreading shit, wouldn't they both be out of business?

But much of what enters the mind as well as what enters the mouth—well, eventually you just have to "let it drop."

Taking it from there, I looked at my watch and realized it wasn't getting any sooner. Reminiscing about the past is good preparation for the future but sometimes it robs the present, so I clicked on to the time of now and took my tail back up the shoulder of the road to check on Mary Teal.

Evidently she was done. This time there were no arms waving me away. Still, I could see her making some kinds of serious motions inside that looked like a priest going through the ritual of an exorcism. Expecting to see bats and crows fly out when I opened the door, instead I only saw Mary Teal continuing with the same motions while adding, "Let me do the cooking tonight. I think you're using too many onions."

Back on the road again, she was driving, I was wingman. The length of the drive, the drone of the motor, and a lack of changing scenery had put us both in silent mode. We were content to cruise along without a word, comfortable with the thought that there are two kinds of time. Time saved and time spent. And yet, since life is mostly made of memories, and memories are value by another name, isn't this where time spent and time saved are one and the same?

Function

"In the art of seeing, the act of looking is one stroke of the brush."

(The artist-philosopher Jyzklerbigev Stlybwuftedevski, largely unknown, because not even his mother could pronounce his name.)

    A Split-window Volkswagen Microbus is a lot like abstract art. Each person looking at a Splittie sees something different. A middle-aged suburbanite might lay eyes on your little German relic and think, "Man, back when I had a life, I actually *had* one of these," while the guy standing next to him is grinding his teeth and trying not to say, "If only I had a hammer—a *big* hammer," because the sight of your Bus brings back memories of the time his daughter ran off with some Bus-driving hippie and came back dirty, drug-addled, and knocked up with twins.

    Brian "Crusty" Holcomb and I got to talking about that. Crusty's a second-generation VW man. His dad Connie has a foreign car repair shop that dates back to the Jurassic, so when Crusty came along it was rumored that he was born with more V and W chromosomes than X and Y, and that during a recent physical examination he was seen to be endowed with metric nuts. They were six-sided and sizeable, according to rumor, and according to additional rumor, Crusty was told by his doctor to take a thirty-three millimeter wrench and twist those nuts either to the right or to the left, depending on whether he wanted his potential son or daughter to vote democrat or republican.

Even without the political considerations, Crusty's make-up and heritage translated into a life-long involvement with fixing and driving VWs, particularly Buses, and man, did it show. He had a sweet-running, showroom-quality camper Bus that was put to use all over the Midwest and took awards at whatever show it was entered. "There's a lot of interest in old Buses," he said. "You really see it at a car show. Your Bus could generate some interest too, if you'd just give it a bath and touch up those rust spots."

Up till then, I never even thought of bringing my Bus to anything other than Bus campouts or VW powwows. Too true to the tribe. But Crusty's words stuck in my ear, especially that part, "if you'd just give it a bath and touch up those rust spots."

Intrigued, I hosed it down, dried it off, and dabbed some paint over the spots of rust. When the paint dried, I took a sheet of 400 grit sandpaper and rubbed the new paint until it matched the dull finish of the old. On the skin of a 40-year-old person, this particular finish would be described as "pre-cancerous," but on the skin of a vehicle with the same age I think the operative word here is "patina."

Patina schmatina, if it wasn't good enough for the car show, then the powers that be could just turn me away.

Which was a distinct possibility because I was tuning it up for *thee car show*. That would be the renowned Iola Car Show, an annual Midwestern event known world-wide and taken very seriously by its hosts and its participants. The show is unimaginably huge, with over 4000 swap spaces, and with easily enough camping area and show grounds to qualify as its own city. The official flyer states that only show-quality vehicles will be admitted, and that cars with rust and cars with doubtful qualifications will be turned away.

Right away there was doubt. First by me, and then by the gatekeepers. As I'm inching the Bus toward the entrance—preceded by a long line of impressively tricked-out foreign and domestic iron, and followed by an equally impressive line of the same pedigree—I began to wonder, "Is it too late to grow a beard and pretend I'm not me." And if not that, then, "Could I just pull off to the side here someplace and crawl under the bed?"

Adding to the sense of insecurity was the feeling that my Bus might end up as food. Jaguars, Barracudas, Vipers, and Cobras were all over the place, and the presence of those kinds of predators gives the driver of something the Germans call a "Kombi" a definite sensation of being on the menu, since "Kombi" is phonetically identical to "combi," which is what you order when you want multiple toppings on your pizza.

To push that sensation out of my head and down through the lower opening of my own personal exhaust pipe, I had to remember that I was there to have fun. I wasn't there to win a trophy so big it needed its own trailer, and I wasn't there to get mad at people who might mistake my Bus for a dumpster. No, I was there to have fun—at least fifteen dollars worth of fun because fifteen dollars was the price of entry.

The fun started when I handed over the money. Maybe I should say the fun backfired when I handed over the money. Because as the line of cars inched their way forward, and as that brought me closer to the shapely blond standing by the gate and taking the money, I also readied my deer leg. Let me explain. People are always looking at the Bus, right? I mean, you stop at a stoplight, and while the thing is red, you notice the other stoppers—especially the ones in the lane directly to your left—are eyeing up the Bus. Now, even though there are plenty of people who have positive feelings about an Olde Split-window Volkswagen Microbus, some assumptive members of society automatically assume that "something is wrong" with a person who would be out in public driving "such an ancient piece of crap." To show sympathy with those members of society, and to bolster their eager assumptions, I will very casually reach up and scratch the side of my head. This with the window slid back and with my head in full view. Only instead of using my middle finger to do the scratching—like my inner imbecile wants me to do—I use the hoof on the end of a deer leg. The leg is positioned within a shirtsleeve, with the cuff attached near the hoof, so when you put your hand inside the sleeve it looks like your hand and part of your arm has been replaced by the foreleg of a whitetail deer. Scratching the side of your head this way is not only fun, but also charitable. You get your tummy tickled, plus you've given a gift to those assumptive members of society who are in the habit of having a profound need to be right. ("I *knew* he was going to do that, Mabel. I mean, look what he's *driving*.")

And just to make it known that I my saintly little self am not above the sin of assumption, I assumed that the shapely young blond guarding the gate and taking the money would not be freaked out by the deer leg. That is, I assumed that when I suited up with the cloven hoof and reached out the window with the filthy lucre tucked between the toes, she would not go bananas. Me, being just your average assumptive member of society, I just kind of took for granted that she'd maybe roll her eyes and say "yeah, right," or at least give off with a giggle. But no, at the sight of the deer leg her eyes got bigger than her own head and she shrieked "Oh my God!" while there I sat, trying to recover by mumbling something like, "Shot off in 'Nam. It's an organic version of the hook."

At that, she snatched the money, ducked and ran, and I took that as my signal to move forward.

Moving forward was like moving forward to another station of the cross.

Only instead of stopping to pray, it was like stopping to get hung on the cross. Two bow-legged old geezers were the problem. Officially, their job was to direct me to the section of the show grounds according to category of vehicle, but when they saw me coming you could tell by the looks on their faces that directing me to a certain section of the show grounds was only going to be a small part of their job. The big part was going to be condemnation. This was especially apparent after taking in the expression on the face of the bow-legged geezer to my right. I had to stop between them, and when I pulled the Bus into the proper position, the best way of describing the look on his face would be to say, "Patton scowls at Hitler."

Since the other bow-legged geezer—the one on the driver's side—was closer and didn't look like he was mentally packing a pearl-handled revolver, I turned my face to him, whereupon he very sternly said, "You a vendor?"

"No, this is a show car."

"A *show* car?" It was Patton. He had an elbow and part of his head stuck through the rider's side window and was eyeing up the inside of the Bus as if it was a moose doot posing as a nutmeg.

Doing my best to lighten him up to where he would no longer think that letting my Bus into the show grounds would signal the collapse of Western Civilization, I gave him my best smile and said, "Well, if you had a classic like this, wouldn't you want to show it?"

There was a long pause and a slow shaking of the head. "I wouldn't *have* something like this," he said, finally, and very reluctantly motioned me to move ahead and follow the line of cars heading for the Post WW II parking area.

The people assigned to direct traffic to the Post War area seemed a lot more sympathetic. None of them were bow-legged geezers posing as military heroes, and instead of glaring at me as I came into view they encouraged me in by waving bright orange batons. As I passed they'd smile and wave again and sometimes give me the peace sign. Nice guys.

Just to be on the safe side though, and just to cover myself in case I was being mildly mocked, when I waved back, I waved back with the deer leg.

The People of Orange Batons arranged us in rows. They directed us along the grass and parked us in row after row after row. I ended up between a sixties Lincoln and a yellow Roadrunner. The Lincoln was about as long as the Land of Lincoln (the whole state of Illinois), and the Roadrunner, jacked up and having a killer finish under many layers of clear-coat, looked just plain cool and fast. As we stepped out of our vehicles, the drivers of both the Lincoln and the Roadrunner tried to look at the Bus and at the same time tried not to look at the Bus. You've seen that look. It's the look of the prudent in the presence of porn—or the look of Baptists trying not to notice one another when trolling the isles of the liquor store.

Up till now, I was the bag lady and the Bus was the shopping cart. Here, among the acres and acres of spit and polish, the others were the polish and I was the spit.

But if you look at the word "spit," all you need is an "l" to make it spell "split." As in Split-window Volkswagen Microbus. That reminder reminded me not to forget to remember that I did not come to the show to compete. The idea was to goof off, and the way you do that with a homely little Splitty in a pretty car show is you kinda sorta poke fun at what the uptight revere. For instance, on most of the show cars was a prominent sign that read, "You can look, but please don't touch." Some carried it even further with a display of multiple signs of the same wording behind a cordon of velvet rope supported on little golden pillars.

Thinking that the show goers might like a little relief from that sort of stiffness, I hung a hand-lettered sign on my outside mirror that read, "You can look, and you can touch, but please don't *shoot!*" On the opposite mirror I hung another sign. It was a full-color picture of a large toad standing high on all four legs and with its mouth wide open. Under the creature were the words, "Microbus parking only. All others will be **Toad.**"

To make people feel even more welcome, or welcomer yet, I modified the windshield placard. The windshield placard passed out to the car show participants read, **Iola Car Show**, which basically begged you to take a magic marker and convert the word "**Iola**" to "**Hola!**" which in French means "free beer."

Once that was done, I went to the back where there was yet another sign. This one was taped to the inside of the motor lid. So with the motor lid swung all the way up into the vertical position, what you were looking at was a sign reading: **Rewind Rope-start. Will demonstrate for beer, flattery, threats, or insults.**

No sooner did I swing the motor lid all the way up than people began to gather. "Oh?" they would say of the little oddity, "Where did that thing come from?" When I explained that it was once part of the starting mechanism on a snowcat, the next question usually was, "Yeahbut, does it work?"

Since it's better to show than tell, and since it's better to participate than be shown, my answer was, "Go ahead. Put your foot up on the bumper. Then grab the handle with both hands and lean back. You don't have to pull hard."

This was always my favorite moment. Groups would gather and disperse, people would crowd around and drift away, but during the whole course of the show, my favorite moment was that short period of time immediately following the invitation, "Go ahead. Grab the handle and lean back."

Because suddenly there was a spotlight. First there was this semi-circle of people, maybe two or three deep, all wondering about the simple performance of a thing. The spotlight is on the thing.

Then, when the closest or the most curious is invited to "grab the handle with both hands," the spotlight has moved to the person.

Now the focus of the crowd has shifted from wondering about the simple performance of a thing to all the complications involving the performance of a person *dealing* with a thing, and the person feels it. Will the spotlight blind or illuminate?

More often than not, it blinds. "No, no, you do it," I'd hear, or, "That's okay, I believe you," as the person is stepping back into the safety of the crowd.

If they themselves were blinded, I myself was blinded even more. What I mean by that is I expected the Big Musclemen to be the most likely to grab the handle and lean back. You know, six-feet two, shoulders like boulders, and arms as big as legs. But when one of those tanned and toned vessels of vanity appeared with a pretty little accessory known as a "girlfriend," he very aloofly demurred, so my next choice was his girlfriend. She took up the challenge, took it up successfully, and when the motor came to life she quickly turned and pumped both fists in the air to acknowledge a huge eruption of roars and cheers coming from a semi-circle of thoroughly entertained onlookers, two and three deep.

All day it was like that. Skinny, nerdy teens stepping up and showing their cautious dads how it was done. Four middle-aged women stumbling in together, half in the bag and laughing, and laughing even louder when one of them said to one of the others, "Hold my beer," and then stepped up, barefoot, and started the motor with one pull.

In the beginning, "Transporter" was one of the formal names given to the Splitty. Back when the Bus was in the first stages of production, German engineers, being German, used the term "Transporter" to convey the idea of utility, and utility only. If you doubted that, all you had to do was look inside. The inside of the early models had little more than three pedals and a gearshift, which left lots of room to transport this and that to here, there, or wherever. It was all about utility.

Little did they know.

Little did anyone know that the stark and utilitarian little box dubbed the "Transporter" would become the vehicle of choice not just for those who wanted to transport cargo, but also for those who wanted to transport themselves out of the ordinary. Because the early Bus was a three-dimensional blank slate. The fact that there was nothing inside was an open invitation to innovate and decorate. With humor and imagination the little Transporters became rolling party Buses, Surf Central, and—love it or hate it—the never-to-be-forgotten "Hippie Van."

In spite of all that, the most common way to be transported out of the ordinary was to rig this box into a camper. As a Bus-driving friend of mine once said, "It's the perfect place to love and keep you when you're not welcome at home." (He's back now, but living in the garage and drinking.) Beyond that, converting a Bus into a camper is all about enjoyment through protection. It's having your own private place of summer while surrounded by winter, or having a warm and dry place to come back to after tramping around in a cold and wet world. This I kept explaining to the car show attendees eyeing up my Bus. If they asked about the wood-burning stove tucked between the driver's seat and the bunk, my ready answer was, "I got tired of waking up damp and cold in a tent and trying to find dry sticks to light with my wet matches." With a stove inside a Bus, you can be serenaded by raindrops instead of feeling threatened by them, and when snow is in the air and frost is on the ground you can wake up impressed with Mother Nature's artwork instead of wishing she would shove it up her ass.

The same with the holes cut in the floor. With holes cut in the floor, gone are the days of hard-water fishing out in the open. No more piling on coats and capes and sitting on a bucket in a bitter wind. Better to sit inside fishing down through the floor, and wearing a tee-shirt instead of a parka, and lounging next to a stove brewing coffee as well as frying freshly caught fish.

There were screen windows to explain, the gluten-free seal-a-meal porta-potty and the a-*maise*-ing organic buttwipe, and the pull-out bunk and lawn chairs, too. There was all that and more, and here is where I was reminded of the lonely Splitty's place in a car show whose main purpose was to feature whole square miles of tricked out foreign and domestic iron. Function. If the singular function of all of the other vehicles came to a halt once the key was turned off, this is where the multiple functions of a Splitty came to life. People walking into the show ambled up and down the rows, impressed with all the parked and dormant muscle, speed, and polish, but when they got to the lonely little Splitty, they stopped. The ambling stopped, and they gathered, laughed, and had questions. Why? Because the little Bus was still active. Parked, it had more function than most other cars have when they are driven. It was as if the other cars, parked, were still photographs, and the Bus, also parked, was an ongoing video showing function and history, some of it funny, some of it instructive, and some of it so far out of the ordinary as to hear people say, "You gonna be here all day? Good. I gotta go find my buddy. He won't believe this."

When the show was done and everyone drove away,

I assumed they were all going home. I also assumed they assumed the same of me. But there was a river along one of the back roads I took on the way to the show, and along the river were quiet places to park. Since one of my hubcaps doubles as a grill, and since I had brats and beer in a cooler in the bus, I pulled into one of the turnouts and set up for the night, reminded once again that a Splitty is not really about show but about function, and the more you have of that, the less you feel a need to impress.

A Blank Slate for the Driver

"Why don't we use Latin anymore? Because it has words like 'Sumus quod sumus' (We are what we are), and nowadays only a fool would be satisfied with that."

(Taken from the first page of *The Fat Sex Fitness Book.*)

    When you drive a Split-window Volkswagen Microbus, people pass you all the time. There you are, bumping along in a little box whose motto might be "Zero to sixty, maybe," and almost as many cars pass you from behind as there are cars coming from the front.

    Rolling along in a vehicle noted for its casual indifference to speed, you too take on this trait, and here you become the driver instead of the driven, unmindful of clocks made of carrots and sticks, knowing that there is value in The Time of Now, and that living in the reality of Here instead of the abstraction of There puts you in a place of no need or desire to match the speed of the perpetually late.

    And so, rolling over a wide road in a big land, piss-emptied and having no hunger or thirst, I had the same smile and the same wave to all the people who passed, regardless if they smiled and waved first or last, or passed fast with middle fingers pumping full blast, aghast at the sass of a dumb ass who would respond to their monodigital insults with an easy smile and a simple wave that included *all five fingers*.

Yup. Physics tells us that for every action there is an equal and opposite reaction, but then there are our mothers. Our mothers, and to a lesser extent our fathers, tell us different. An equal and opposite reaction? For the vengeful, maybe, but woe to the vengeful, and pity them, for their memories are incomplete.

Which probably means to blissfully dismiss the insults of others, and very amiably respond to their words and deeds by turning the other cheek, a lower cheek, and with the thought that they may kiss it.

Except when it comes to the "f-word." When someone passes you, points a finger and yells the dreaded f-word, is this a time to respond with only a nod and a smile?

You be the judge.

It was a cool day in summer and the road was wide, smooth, and straight. The Bus was moving along at a thrifty fifty in a no-wind situation when all at once here comes a station wagon. It's coming up from behind, and fast. When it very quickly went to passing mode and veered into the oncoming lane, I noticed a woman in the passenger seat with her head out of the window. Her arm was out too, and seeing that, I got ready to flinch, since past experience told me I was about to be "gifted," and by "gifted" I mean hit by an egg, a beer can, or a paper cup full of coke and ice left over from a greasy meal at a fast food joint.

Turned out it was worse than that. She hit me with the f-word. Even before the station wagon pulled parallel with the Bus, she was yelling "fire!" and for added emphasis was pumping her arm and pointing to the back of the Bus in the manner of a republican pointing at a democrat when the democrat wants to waste your tax money on welfare for the poor instead of wasting it on welfare for the rich.

Politics aside, this was no time for a half-fast response. Instinctively I slammed on the brakes and jerked the Bus over to the shoulder of the road. Almost before it stopped I was out the door and at the back where clouds of smoke were billowing out in four different directions.

A quick lift of the motor lid revealed the source of the smoke. It was the engine. The engine was on fire. Dirty flames were leaping from the spark plug wires, and even dirtier flames were flaring up from the fuel line.

The fuel line was what put me in motion. Made of rubber, the fuel line was likely to burn through at any second, and once it did, a whole tank of gas would be let loose and you know what would happen from there, so I fell to the shoulder of the road and grabbed handfuls and armfuls of dirt and gravel and kept heaving it into the engine compartment until all that was left of the fire was thin wisps of dying smoke seeping up through the gritty mess.

When my heartbeat dropped from a thousand beats per minute down to around five or six hundred, I started to wonder what happened. Duh. You had a fire, Sherlock Holmboy. Yeahbut, what caused it? The day was cool, the driving was easy, and the fan belt was not frayed or broken.

Oops. The dipstick rag. To check the oil, I had a red shop rag crammed into a small circular cutout in a sheet metal brace just inside the engine compartment. The rag was gone. Just where the rag went was no longer in doubt when I reached over the top of the engine and into the fan housing and felt cloth instead of metal.

Great. I just fried a perfectly good motor by violating the rule that warns the driver to keep the engine compartment clean. No paper, no rags, no nothing that can get sucked into the fan and choke off the stream of air that keeps the temperature of the motor below the critical stage of Spontaneous Volkswagen Combustion.

After several minutes of mentally kicking myself, and after a long time of wondering why a merciful God did not equip time with a delete button, I did the sensible thing and looked around for someone to sue. Okay, not really, but as long as I was in a big snit about having lost *yet another* motor to the effects of heat (Full disclosure: yes, this was a repeat of similar situations), finding someone else to blame after this most recent episode of personal failure—well, isn't that the American way? Anyway, the first place I looked was away from America and over to Germany, and why not? Those krauts over there, they were just as responsible for this most recent fire as I was. Oh? Yes, because they built me a Bus whose motor was not equipped with a temperature gauge. Think about it. Without a temperature gauge, how am I supposed to know if my motor is suffering from a heat-related problem? And here you might rightfully say, Well, Dipstick, if you'da kept the oil rag out of the engine compartment, then you wouldn'ta *had* no heat-related problem. And to that I'd answer, Yeahbut, what about the countless other causes of excess heat? Like for instance a faulty thermostat, which I've had, a broken fan belt, which I've had, or too much advance on the timing? Which I've also had. I could go on and on, but the question remains: Without a temperature gauge, how is the driver to know if the motor is suffering from a heat-related problem?

First of all, what a dumb question. Because all you can get from a question like that is a passive answer when what you really need is an active solution. Going from there then, the question should be: Where can I get me a temperature gauge?

By then there were enough drivers of early Buses to support what is known as an "aftermarket." An aftermarket supplies parts for an original purchase. These parts can match the parts of the original purchase or they can provide an "improvement." So if you drive a Split-window Volkswagen Microbus and you're tired of being blind to the plight of your struggling little forty-horse power source, what you do is you go to one of the aftermarket suppliers and come away with a little box. Inside you find a sensor, a wire, and a little round gauge with a white face and a black needle.

Here's where I decided to make a deal with the devil. Let me explain. Some owners of early Buses are seen as "Purists," and have talked themselves into believing that the Split-window Volkswagen Microbus is a sacred object. Moving, it is a rolling icon of automotive perfection. Parked, it is a shrine. For you to disagree by saying the Bus is really kind of a blank slate for the driver to add to according to his own personal needs is to be branded a heretic. You do not alter or add to such a divine creation. This is an act of sacrilege, and thus was born a mutant mixture of philosophy and superstition, one of whose tenets is: "Drive thy Bus, O thee, and do not defile with schemes to make good into better, for good is happy with itself, and what is happy must rightly remain the same." To that you can add either "amen" or "excrement of the he-cow," depending on whether you think VW stands for "Veteran Wanderer" or "Vacant Wacko."

Once I popped the temperature gauge into a neatly cut hole in the dash, and once I ran the wire to the motor and bolted on the sensor, it was as if I had grown a nerve. To start the motor and to watch the needle move up and hover in the ideal operating range gave a sense of quiet satisfaction. To see the same needle climb into the red zone caused pain. If the needle did not climb at all, this was a cause for worry. No doubt about it. I felt bodily connected to the motor. Yes, the gauge, the sensor, and the wire were three additions to the Bus, but beyond that, there was an addition to me, and it had more to do with knowledge than with numbers.

The sense of gain didn't stop there. The little round gauge with the white face and the black needle was like a quiet witness called on to testify against loud forms of belief. Which is a good thing, because belief, especially the loud form, can be dangerous. If you look at that kind of belief, it's like contemplating the properties of gas. Gas fills an area much quicker than something solid, like truth. So if you have empty spots in your head—which would be equivalent to questions—then gas is likely to fill them first. It flows in and takes up all the room, and exists as belief. If something solid intrudes, like truth, what does belief do? Does it accommodate its elder brother? No. The belief behaves like gas. Forced into a smaller space, it heats up under the pressure, and the pressure puts up resistance to the entrance of any more solids. Unless the gas can be vented in favor of letting in something more substantial, the heat and pressure get to be too much, and the resulting explosion leaves the solid part in pieces and allows the gas to go off and find another empty space. Whole civilizations have fallen victim to this phenomenon, and another one will fall soon, but before it can get that far, individuals have to be infected—intelligent people like Us, and stupid people like Them.

With me, the belief was that the little air-cooled motor living in the back of the Bus kinda sorta took care of itself. Once you filled the tank and topped off the oil, German engineering took over from there. Oh, you still had to periodically set the timing and adjust the valves, but what I'm saying is, my belief was that Teutonic technology pretty much had me covered. The people in the Wolfsburg factory did their job by building a Bus that worked, and that let me do my job, which was to drive it, park it, and drink beer, laugh, and fart.

But then the gauge. Once the little round gauge with the white face and the black needle was added to the Bus, silently, truthfully, it added to me. For every belief it affirmed, another it destroyed. It told me I was right to think that the temperature of the motor stayed within its operating range while the Bus was moving along on a cool summer night at fifty-miles-purr, but if I was driving on a hot day at the same speed and thought the motor temperature would remain the same, here's where the gauge told me to think again. Even though it's not polite to point, the needle pointed at a higher number. The number was only about 40 degrees up the scale, but that number shot up another 60 clicks when pushing the same Bus on the same kind of day into a strong headwind.

Which meant that the motor was laboring along at 100 degrees above its ideal operating range.

Not good, and here's why. Excess heat can ruin a perfectly good bought-and-paid-for VW engine in only a little more time than it takes a skinhead to comb his hair. Even if the temperature of the motor does not rise to the point of self-ignition, you have to remember that metal expands when tested by heat. Some metals, like some waistlines, have a tendency to expand more than others. For instance, the people who study physics tell us that aluminum expands many times more than steel when both are heated to the same degree. Volkswagen made their cylinder heads out of aluminum. Think of the cylinder heads as the top half of the motor. The bottom half—the engine block—is made of magnesium. Steel bolts anchored in the magnesium block extend to the heads to hold the little power plant together. The good news is that both aluminum and magnesium are light in weight, which means more power per pound, and which makes for easier throwing during a temper tampon. The bad news however, is that both magnesium and aluminum are soft metals, and can be whittled like wood. Overheated they are softer yet, and when you combine that buttery quality of the metal with the capacity of aluminum to expand more than the steel that holds it in place, what you have is the expanding aluminum cylinder heads literally pulling the non-expanding steel bolts loose from the soft magnesium block. Later, when the motor cools, the aluminum shrinks back to its original size, and this makes gaps where once there were seals, and since oil is more loyal to gravity than to a crankcase, much of that oil lets itself out through the gaps to lubricate the roadway, while the rest slimes the exterior of the engine, collects dust, and the dust and the oil acting together form a coat of insulation to further compromise the capacity of the engine to cool. Yes, the motor still runs, but the poor bugger done had a near-death experience and has suffered a permanent injury. Just one more example of what happens when a movable object runs up against a resistible force.

The way to keep my next motor from falling victim to yet another resistible force was almost too easy.

Just give it more air. Since the motor was air-cooled, why not? If giving it less air made it hotter, and cutting off all air set it on fire, then it stands to reason that giving it more air will cool it down.

For this, I decided to use side scoops. Side scoops are nothing more than sheet metal add-ons bent in a way to funnel more air into the engine compartment. Fastened to the outside of the Bus near the back where the air is drawn in, they can be made to look "stylish" if done right, or, if you have nothing but contempt for all things fashionable, they can be made to look like cauliflower ears on someone whose head is shaped like the water closet on the back of a flush toilet.

My first attempt at making side scoops ended up looking like something in between. Let's just say they were esthetically acceptable when seen from the moon, but when you moved in closer they each had that look of a dustpan that got caught fooling around with the wife of a hammer.

Regardless of their appearance, their effect was eye-opening, and all verified by the gauge. Hot weather and headwinds no longer put the motor in a fever. In fact the side scoops were self-regulating. The faster the drive, the more the scoops funneled in air to the fan. Driving into a headwind funneled in more air yet, and that extra volume whisked away the excess heat that would normally build up if the cooling fan had to suck in air instead of having it served under pressure. Bottom line: Not only did the gauge verify that the side scoops nullified the motor-melting effects of driving into hot-weather headwinds, but those same scoops in ideal driving conditions kept the temperature of the motor on the cool side of its normal operating range. All good.

Still, even with those heavenly results, the Purists thought of it as the work of the Devil. Putting on side scoops fell under the heading of Original Sin, which meant to alter or add to the original Bus was a sin. I could hear their dogma barking in my ear. "Put thy faith in what is done, Demon Son, and challenge all that is to come."

But the needle on the gauge, unaware of the True Believers and their need to be right, silently and truthfully testified that Belief, like gas in a balloon, can lift the spirit, but without the backing of its elder brother, Truth, there is always danger of a fatal plunge.

Going from there, the gauge also put me wise to the fact that cold could be as damaging to the motor as heat. The problem was, in cold weather the motor was too slow to warm up. Say you wake up on a winter morning to let the dog out, and when he lifts his leg against a snowbank the little yellow stream doesn't even melt a hole. So what? you say, this is the upper Midwest. In winter we have cold here, and lots of it, but vut da heck, as dey say in Deutschland, if you need to go anywhere all you have to do is step into the Bus and flip the key, and the motor, being German, obeys. It starts, and runs, and runs without complaint.

Something, however, is wrong. Even if the motor runs, and runs without protest, here's where the needle on the gauge points at more than only a number. It also points at a problem. By not sweeping up and hovering in the normal operating range, the needle bluntly states that the motor is too cold to properly atomize the gasoline. It's saying that some of the gasoline sucked into the cylinders during the intake stroke is not entering as vapor. It's entering as liquid. Believe it or not, liquid gasoline does not burn. Really. Look it up. Only the vapor burns, so some of the liquid, unburned, gets whooshed out the tailpipe, and the rest gets blown past the piston rings during the power stroke. Bad move, because whatever gets blown past the piston rings during the power stroke ends up in the crankcase.

And the crankcase, as we all know unless we don't, holds the oil, and oil is what lubricates the engine. (As if you didn't know that.) But to do its job, oil has to stay oil. That is, it has to stay slippery. However, if gasoline is added to the oil—which happens when unburned fuel is blown past the piston rings during the power stroke—gradually the oil gets more grabby than slippery, and if this process is allowed to continue, the oil goes from a lubricant to a solvent and eventually has the chops to grind your motor to a screeching halt.

The Purists get philosophical about this and paraphrase Ecclesiastes by saying, "To everything there is a seiz'n," but a better approach might be to set aside some of your beer-drinking time and use it for the purpose of installing of a Midwestern Midwinter Thermostat.

You never heard of a Midwestern Midwinter Thermostat? Me neither until one day finding it lying in pieces in different parts of the same junkyard. Strange how that works. I'm poking along past piles and boxes of junk, and what do you know. Here's a lawn mower throttle cable here, and over there sits a roller mechanism from a file cabinet, and didn't I just walk past a length of light-gauge channel iron? Yes, in fact, I did. So I collected those three items, and they, along with a small square of thin plywood, some screws and braces—all that ended up as a Midwestern Midwinter

Thermostat. It was only a matter of putting those pieces together so that a pull on the lawn mower cable in the front closed off the cooling air to the fan in the back, and then a push on the same cable did the opposite.

But why would anyone in their right mind feel a need for such a contraption? I mean, didn't the motor come *equipped* with a thermostat?

As a matter of fact it did, but the thermostat put on at the factory had a wait-and-see attitude. When called upon to open or close, it seemed to say, "Hey wait a minute. I'm a German, not a Russian, so let's not go Russian into things." This condition was made worse by the flaps. The flaps controlled by the thermostat were too restrictive in hot weather driving, and in cold weather driving those same flaps were insufficiently closed. In moderate weather, such a setup generally kept the motor happy. During cold snaps though—like in weather where you have to break up your snot with an ice pick before you can blow your nose—here's where the limitations of the thermostat put on at the factory have the motor feeling just like you would feel if you found yourself naked on a cold night trying to sleep outside in wet grass.

To alleviate the suffering of the forty little horses shivering in the back, I unbolted the thermostat and pulled out all the flaps, and when the Midwestern Midwinter Thermostat was in its place, bringing up the temperature quickly was as easy as giving a pull on the cable, and regulation of the temperature was only a matter of pushing on the cable until the needle on the gauge pointed at the number where the motor ran best. Now the motor had the heat it needed to vaporize the incoming gasoline, and now that gasoline could be used to power the motor rather than poison the oil.

There was more. Even though the collection of junkyard parts solved the problem of cold weather oil pollution, liquid gasoline was still sneaking in through another channel. I say, "sneaking in" because it was doing its dirty work totally under cover and out of sight, and when I finally found the source and the method, it was like catching a dieter getting up in the middle of the night to raid the refrigerator while you yourself were only getting up to stumble into the john and download the wagpipe.

The discovery started by cutting another hole. When all the Purists were asleep or out of town, I made a cut in the sheet metal covering the top of the engine compartment. Normally, the engine compartment is only accessible from the rear, but if you cut a rectangular hole, say, 15x22 inches in the sheet metal directly over the motor, this gives you access from the top. Now, instead of having to pull the whole motor completely out of the entire Bus just to do some simple little fan or generator work, all you have to do is lift the cover off your recently cut hole, and there everything is—fan, generator, spark plugs—all that and more, and visible and get-at-able from the top.

Oh. And the carburetor too. In fact, the carburetor is the first thing you see when you lift off the cover and unhook the air cleaner. There it is, that little dispenser of air and fuel, bolted to the manifold, and plainly visible from the top. Why is viewing it from the top so important? Because if it's only visible from the side, you can't look down the throat to watch the inner doings when you work the accelerator lever. You can only speculate. Who knows? There could be a right, vast-wing conspiracy going on in there.

Turned out it was even worse. It was a plot to further

poison the oil, and it was immediately seen that the plot was being perpetrated by the accelerator pump.

The accelerator pump is built into the side of the carburetor, and its stated purpose is to give the motor an extra little squirt of go-juice when you step on the gas. The motor needs this extra little drink when called upon to pick up speed in the same way you need an extra little drink when called upon to do your taxes.

But wait a minute. When I looked down into the throat of the carburetor and pulled the lever on the throttle to activate the accelerator pump, the nozzle shot out a solid stream of raw gas. Not a mist or a spray, like you'd think. No. A solid stream of raw gas, and lots of it. I mean, if this was pure ethanol instead of only a blend, and say you held a shot glass under the nozzle after three pulls on the throttle, what you and your buddies would have is a good start on a night of pure and simple *fun*.

Instead, it was sobering. Truly this was a problem of ethnic magnitude—meaning the German who designed the accelerator pump should be kicked out of the gene pool.

But problems, as Harry Spencer used to say, are only solutions in the formative stage. Taking it from there, anything in the formative stage needs to be nourished. And what could be a better nutrient than thought, a vital building block in the development of action?

Thinking about the solid stream of raw gas pollutedly shooting down through the throat of the carburetor stayed with me all morning and most of the afternoon. Not until later in the day did the solution present itself, and it happened during one of those cleansing rituals that should be performed at least once a year—or maybe even once a month if you do all your own Bus maintenance. What we're talking about here is a hot, soapy shower.

This particular shower though, turned out to be more than your usual body scrub-down. Much more, because along with the soapy removal of driveway dirt, ground in gear oil, and gooey-gummy tranny boogers, there came an opportunity to instruct and to be instructed—all in the same moment.

The opportunity presented itself in the form of my feisty little housemate, locally known as Pop Can Tessie, picking up the hose outside to water the flowers. Now let it be known that this urban campsite of ours—listed on the tax rolls as a "house"—has the cleansing cave on an outside wall, and that wall has a window which allowed me, showering inside, to track the movements of my feisty little housemate outside. So not only could I see her pick up the hose, but I could also throw out a comment, since it was summer, and the only thing separating my big mouth from my hose-toting housemate was a screen.

(Let us now create a space in silent commemoration of what can happen when someone with a big mouth tests the nerves of someone armed with a deadly hose.)

[space]

Okay, so you know what's coming. But what's coming is only the beginning of the story. Maybe this particular beginning will make you laugh if you're a member of the Guy Persuasion. If you're a Person of Femality, though—especially a Person of Femality who feels a frustrating need to civilize members of the Guy Persuasion—this particular beginning might have you quite justly oozing with malicious satisfaction, which will be commemorated here with four lines of verse:

> Yes I chose
> without a hitch,
> to turn the hose
> on the sonofagun.

So yeah, she got me big time. The water was cold and had me yelping like a wounded puppy. Ironically, the comment that caused it was entirely uttered for her own good. You decide. After she picked up the hose *and* lit a cigarette, I said, "If you really want to do something good with that hose, take it off the flowers and douse that smoke."

Anyway, laugh if you want, or ooze with malicious satisfaction, but save some of your focus for what really counts. Because along with me yelping like a wounded puppy when hit with the hose, I noticed something. I noticed that the water from the outside of the screen came at me in a solid stream. But going through the screen, that same stream was broken up into a spray and a mist. Let it not be said that a common road rummy can't notice and yelp at the same time.

The resulting sense of discovery had me quickly drying off so I could go back out in the garage to get dirty again, and there, in the garage, in less time than it takes a carpenter to affirm that yes, Pinocchio's lower back *does* have lumber vertebrae, on my hands was a good start on another month's worth of accumulated dirt, and *in* my hands were two carburetor gaskets with a piece of window screen tucked in between. The dirt I could ignore until June—this was May—but the gaskets and screen deserved immediate attention. Because if a solid stream of water could be turned into spray and mist by passing through a screen, then the same should happen to a stream of gasoline, since even this seventh grade dropout knows that water and gasoline are basically the same except one is a little more volatile than the other because, well, one comes from a more volatile part of the world than the other. (Duh.)

You can see where this is going.

First the gaskets and the screen were going between the carburetor and the manifold. Then I was going for a ride. It would be a long ride, and during stops I would routinely check the oil. Pretty soon I'm seeing cleaner oil on the dipstick. If the dipstick formerly showed blackened oil after 500 miles of driving, now, with the screen in place, the golden liquid is taking longer to get dirty. Not much longer at first, but over the course of many miles, and over the course of boocoo oil changes, the improvement was undeniable. Instead of the oil going black at 500 miles, at 900 miles the dipstick would show a color best described as "caramel." (Not to be confused with "camel." Camel is an odor, not a color.)

So what's at work here? Well, if you're a conservative, the obvious answer is the government finally got out of the way and everything started operating more efficiently. Or, if you're a liberal, your answer would be, "It's not God that works in mysterious ways, but the magic of the federal bureaucracy." Don't even bother to ask the followers of Hate Media because, celestially speaking, these people are more inclined to think up Uranus than down to Earth, and in such a state have no agenda beyond making a big stink.

But if you were born in the Year of the Bus, and if you were baptized in a parts-cleaner tank, you instinctively know that what's at work here is the motor has gone into detox, courtesy of the screen. With the screen breaking a solid stream of gasoline into droplets and breaking the droplets into mist and vapor, the motor is having a cleaner burn. And without the contamination caused by piston ring blow-by, the oil is having a longer life. Which kept getting longer because a continued cleaner burn sweeping through the combustion chambers allowed the oil living in the crankcase to gradually wash away built-up crud down there—crud caused by formerly dirty burns.

It wasn't over. In fact, the best was yet to come, because like herpes, this wire screen thingy turned out to be a gift that kept on giving. If it gave more life to the oil, it gave even more life to the gasoline. Since the screen, along with the Midwestern Midwinter Thermostat, was giving the motor a cleaner burn, this naturally translated into more miles per gallon. Running the numbers showed an improvement between 4 and 16% better gas mileage. The discrepancy between 4 and 16 depended on how you drove and where you drove.

Intrigued, I added another screen. This one was rotated 45 degrees below the first. That is, if the wires on the top screen were pointing up and down and left and right, the wires on the lower screen were rotated 45 degrees off the perpendicular. This means that any droplets not broken up by the top wires would run smack into the offset wires below. But not *directly* below, since the screens had a spacer between them so as to break up the flow without restricting it. The thinking here was if one screen gave 16% better mileage, then two would give 32, four 64, and so on until the Bus was actually making gas and I could create gasoline just by driving, thereby saving the world by having the capacity to send the excess back to the Middle East to be re-pumped into the ground. (Don't be foo, stay in skoo.)

When adding the second screen only gave slightly better mileage at slower speeds, I figured my next best chance to save the world was by making an appointment to undergo a vast ectomy.

Even without planet-saving implications, the improvement in mileage was enough to urge me into giving the little gas-saver a name. Calling it "The Little Wire Screen Thingy" seemed about right when I was kicked back in my underwear and tipping a beer, but after the beer was gone, and after I dressed a little more formally, I renamed it the "Cyclonic Vacuum Diffuser." You can see why. No, it's not because engine vacuum pulling air through the throat of the carburetor ramped up velocities that were downright cyclonic and thereby diffused the droplets of gasoline whipping through the screens. No, it's not that at all—even though that would be a good excuse. Why I chose "Diffuser" over "Thingy" was because "Thingy" sounds like it came from the mind of an idiot, whereas a mind that uses words like "Diffuser" is likely a mind that has risen above the designation of an idiot, passed the level of an imbecile, and has been lifted into the exalted state of a retard. So if I could talk myself into believing I had reached the enlightened level of a retard instead of existing as a lowly idiot, then I could also talk myself into canceling that appointment for a vast ectomy.

Which calls up the question: Is knowing the limitations of your Bus and the limitations of your self the first step in a plan to exceed them, or is it only the first step in a scheme to deceive them?

End of Report

"The job of the police is to protect me from the person they think I am."

(The entire text of an acceptance speech given by the winner of a Charles Manson look-alike contest.)

When the truck moved in closer, I decided to get serious. Just in case. It was a pickup truck, old, and had been following me for the last few miles, first at a reasonable distance, and then moving in so that less than a dozen yards separated the front bumper of the pickup from the back bumper of the Bus.

Since this was a reminder of a time when a similar-looking pickup closed the gap between the bumpers and deliberately pushed the Bus down the road and put it in pieces in the ditch, I figured, well, time to prevent history from repeating itself, and the way you do that is you load up the little bone buster that goes bang when you pull the trigger.

Specifically, the little bone buster was a .45 automatic, the same kind of gun I carried while in uniform. A .45 always works. "Dirty, wet, or neglected, rack the slide of a .45 and you're protected," was our motto back then, and to that I added later, "Drive forty-five, pack a .45, arrive alive."

It always works.

In two easy moves, I had the semi-loaded gun on the

floor next to the driver's seat. I say semi-loaded because the clip was only resting in the slot. The gun was upside-down in a partly opened case and had the clip extending halfway out of the grip. Look at it this way. This way the gun is super-safe, but at the first sign of hostile action you can reach down with one hand and in one motion push the clip into place and lift out the gun. Or, if it turns out that there's no problem, then it's only a matter of plucking the clip from the grip, popping it back in the pouch provided, and zipping up the case. Case closed.

But the pickup persisted. It stayed close as if to pass. Odd, because we were traveling on a straight road in a part of Illinois that defined the word "flat," so that fact, along with the fact that there was no other traffic, oncoming or otherwise, kept me alert to the point where the scene in the rear-view mirror had me more occupied than the sight of the road ahead.

Even at that, the road ahead was where the game changed. It started with a curve, and the curve gradually gave way to a hill.

But not a natural hill. This hill was man-made and was put there as an approach to a bridge. It was a new bridge and a high bridge, and as I came around the curve and made the climb, off to my left and down below I could see what was left of an old bridge. The old bridge had been built much lower than the new one, and apparently had suffered a washout, since one side was collapsed and was covered with whole trees and parts of other trees.

Also, down below and off to the left was a road. It was the old road leading to the old bridge. The road was cracked and heaved and had tall weeds growing up through the cracks, but as I crossed the bridge and came down the other side, I could see that this part of the old road was approachable from the new road.

Here's where everything went from bad to good and then back to bad again before ending up as better.

A quick glance in the rear-view mirror told me that nothing in the back had changed. The pickup was still uncomfortably close, so without putting on my turn signal, I suddenly braked and made a sharp turn that quickly took me off the new road and put me in the opposite direction on the old.

I didn't stop there. I kept on bumping along toward the old washed-out bridge, and kept at a speed safe enough to negotiate the rickety road ahead while still keeping an eye on developments behind.

The pickup kept going, but just barely. It appeared to be coasting. Here's where I thought, "If he stops and turns around, that's when I push the clip all the way down in the slot and rack the slide."

Not that I was up for shooting someone. It's just that the real purpose of packing heat is that by packing it nobody has to get hurt. As contradictory as that sounds, you have to remember that a gun is at its best when used for defense, and people are at their worst when on the attack. Now, if someone is on the attack and you have a gun, what are you going to do? Are you going to shoot? Don't. First you want to show. You very firmly want to show that you are not willing to be rolled over, and the way you do that is by standing tall and standing still while holding a gun. Likely result? End of attack and nobody got hurt. To find the likely result if you were unarmed, open up any newspaper and read about the victims in the crime report.

By then I was up near the old bridge and had come

to a stop. The pickup was farther along yet, and since the bounce from the road no longer impaired my view, I was relieved to see the truck gradually pick up speed and finally disappear into the distance.

Good. And good riddance.

Even though he was no longer on the radar, he was still somewhere. Maybe he was waiting somewhere. They say waiting makes it better, so if he was waiting somewhere, to help make it better for him, I decided to stay parked for a while and use the time to take a little breakfast break. It was now or never. I had to act soon or not at all because it was late in the morning and if I held off any longer I was likely to miss out on a good breakfast by having to call it lunch.

Scotty was in full agreement. The whole time of being followed—including the sharp turn and the bumpy ride up to the bridge—Scotty spent stretched out on the bunk in the back, oblivious to any threat. But once I pulled out the pots and the pans, he climbed down, stretched, and looked at me as if to say, "Did you call?"

At that I asked him what he'd like for breakfast, eggs over easy with a side order of bacon and hash browns, or hash browns and bacon with a side order of eggs? "You're going to spoil that dog," people say when they see us share big plates of food looking like something from the menu of a five-star greasy spoon, and to that I say, "Why not?" A dog ain't a kid. Kids, you have to grow them up to go out on their own. They won't do that if you spoil them. A dog though, you don't urge him out the door to go pursue a career—unless it's off to Washington to guide the blind—because his thing is to be your best friend. Taking it from there, when you invite your best friend in for dinner, do you feed that person the same thing you cooked for yourself or do you serve him up a sloppy gob of Gravy Terrain?

No sooner did I step outside to put Scotty's plate on the ground when here comes this big screech of tires out on the highway. Looking up, I saw it was a cop car. Two cops were inside. The driver had slammed on the brakes and was quickly backing up. The car had passed the intersection of the old road and the new road, and now was quickly backing up to where the roads connected.

Before I could wonder why they were Russian around when this was America, the driver wheeled to the right and roared and bounced up the rickety road towards me and Scotty. About thirty feet from where we stood, once again the car came to a screeching halt.

Almost at the same time, the two cops got out and stood behind the open doors. One of them called out, "That dog bite?"

"Not yet," I wanted to say, but instead said, "No, he's harmless—like me."

Suspiciously, they advanced. They came out from behind the protective doors and alternately eyed up first me and then Scotty. When Scotty left his breakfast plate and ambled over towards them with not just his tail but his whole body wagging, one of the cops put his hand on his holster, and me, not wanting it to go any further than that, I said, "Don't worry about him. He thinks you're the mailmen. His birthday is tomorrow and he thinks you're bringing him a present too big for one guy to lift."

At least that got him to take that hand off that holster. Still though, you couldn't help but feel the tension. Something was up. The quick but cautious approach, the continuing suspicion—I started to wonder if it had anything to do with somebody in an old pickup, but before I could ask, the cop who had only recently taken his hand off his holster bluntly demanded as more of a statement than a question, "What are you doing here."

Since I was standing in one place with a plate of food in my hand, and since Scotty had gone back to his own plate, it seemed pretty obvious that what we were doing was eating. So to answer the question as briefly as possible and to show that I had no intention to deceive, my simple answer was, "Eating."

Somehow this wasn't good enough. One of the cops moved in close and stood by me and the other one took a few more steps forward and looked into the Bus. When he saw the fishing rods clipped to the ceiling, he said, "Are you fishing?"

"No," I said. "Just eating."

That still wasn't good enough. He glanced over at the washed-out bridge, and probably assuming that there was water below, put the question to me again.

Apparently the answer, "Eating," and "No, just eating" needed some elaboration, so I said, "Feel the lines. Reach up and feel if the lines are wet on the spools. If they're dry, that means I haven't been fishing."

Ignoring my suggestion, he went on with, "If you're not fishing, then why the fishing gear?"

"We were fishing in Missouri."

"Do you have a current valid fishing license for the state of Illinois?"

"No, we're just passing through. We're coming from eastern Missouri and going to northern Wisconsin, and the quickest way to get there is by taking a left turn through Illinois."

"Do you have any fish with you now?"

"Not anymore." I licked my lips and reached down and patted Scotty to let them know that I was speaking of fish in the past tense.

"You do know, don't you, that possession of game fish in the state of Illinois without a current valid Illinois fishing license is in violation of the law. Whether or not you caught them here."

I shook my head at that, partly for the fact that I did not know there was such a law, but mostly for the fact that the people who wrote and passed such a law should have their heads candled.

He persisted. He certainly was a stickler. While the other cop maintained his position in front of me— him I began to think of as Officer Standby—Officer Stickler kept poking around. "Mind if I look in the cooler," he said. I say, "he said" rather than "he asked" because most of what came out of his mouth came out in "cop monotone," that particular kind of police speech that sounds more robotic than human.

Trying not to sound irritated with it all, I gave it my best shrug and said, "Go ahead. Just don't leave the top open. I'm almost out of ice."

"Have you been drinking," he said—this after seeing that the cooler did not lack for beer.

Resisting the urge to say, "No, but as soon as you two dinks are gone I'm going to pop one in celebration, I kept it short by only saying, "No."

"I need to see a driver's license."

When I pulled the license out of my wallet, he took more than the usual time to look it over, and when he handed it back to me he said, "Can you provide proof that the dog is up to date on his vaccinations."

"He's good. Check his collar. There's a tag."

Instead of checking the tag, the cop turned back to the Bus.

And that's when I realized it was all over for me. Because his next question was, "Are there any firearms in the vehicle."

"Yuck!" I almost said—or something that rhymed with yuck. Here I'd forgotten all about the .45. When I pictured it still upside down in the partly opened case and having the clip halfway down the grip, it was like getting slapped in the face with the tail end of a dead fish. Great, just great. Here these guys are failing to nail me with niggling trivialities, and then along comes me, presenting them with the Holy Grail, and at that, all I could say in answer to the question was a very resigned "Yes."

"Where is it."

I waved my hand. "Between the front seats."

"Is it loaded."

"No," I said, knowing that "Not quite" would be an unacceptable answer, and that "Yes" would have me immediately in handcuffs.

The closest reach to the area between the front seats was through the rider's side window. Since it was a warm day, and since the window was open, Officer Stickler put his head and his arm through the window to retrieve the gun while Officer Standby kept on guard.

But then something happened. When Officer Stickler withdrew his arm, his elbow hit my dog tags. I had my old military dog tags hanging from the grab handle on the dash, and just by chance his elbow brushed against them on the way out and the resulting jangle caught his attention.

"These yours?" he asked, stopping, and looking up from the tags.

When I nodded, he looked back at the tags and said, "What's your service number."

You never forget your service number. In boot camp they brand it on the backs of your eyelids in the same way it's stamped on the front of your dog tags.

When I rattled off the seven digits, Officer Stickler was standing outside the Bus with my gun in his hand. He looked in at the dog tags and then back out at the gun. Shaking his head, he said, "I don't know how the law would see this in your state, but here in Illinois we're talking a ten thousand dollar fine and three years in jail."

When he added that the vehicle would be impounded and sold at auction, and that the dog would have to be given up or be euthanized, that's when I decided I'd rather be dead than lose my dog and my Bus.

First though, I wanted the police report to state why "the subject" put up a fatal struggle. To keep the report from only stating that "the subject became agitated during the arrest and the resulting altercation led to his death," I wanted it to read:

When [Officer Stickler] attempted to handcuff the subject as subsequent to the arrest, the subject became agitated, stating (quote), "Doesn't it just grind your gums that a guy can give up more than four years of his life *defending a whole country* in an outfit trained to shoot guns firing *bullets as big as your head* only to be denied the right to carry a little pop-gun back home to defend his sweet little *self?"* The subject then pushed past [Officer Stickler] and myself, and reached into his vehicle with his right hand. Due to the continued agitated state of the subject, and thinking that the subject was reaching for a weapon, both [Officer Stickler] and myself drew our weapons and ordered the subject to the ground. Ignoring our command, the subject became even more agitated, withdrew his arm from the vehicle and threw the contents of his hand at myself and [Officer Stickler]. While this turned out to only be a set of military dog tags, at the time [Officer Stickler] and myself assumed it to be a knife or similar weapon of lethal design, and responded with proper police procedure by each firing one shot, with the shots striking the subject in the head and upper chest. When the subject fell back and collapsed, the subject's dog attacked [Officer Stickler] whereupon once again I responded with proper police procedure by firing another shot which killed the animal and kept [Officer Stickler] from intended harm.

A copy of this paper and a copy of the post mortem will be sent to the evidence locker, along with the three spent cartridges and the military dog tags.

End of report? Wait.

Because before it got to that, Officer Stickler had one more question. Dropping the pattern of "cop monotone" and letting his voice reflect an attitude of genuine inquiry, he looked at my .45 and asked, "Do you mind if I ask why you feel a need to carry such a weapon?"

No, I didn't mind, and I started by telling that twice I

had been the intended victim of people bent on murder—not counting service time. The first time I escaped with a flurry of feet and fists and the second time I was saved by what could only be called dumb luck. So that's twice. If you want to compare life with baseball, three strikes and you're out, so I decided to face the next pitch with the same gun I carried in the military, a big, bone-busting 1911 style .45 automatic. Those things always work, so when that pickup kept following me at a threatening distance, I knew I was covered. If history wanted to repeat itself, then my best chance at intervention lay upside down in an open case and ready to accept an 8-shot clip.

Here's where Officer Stickler once again looked in at the dog tags and once again looked down at the gun. With a big sigh and a shake of the head he pulled the clip from the butt of the gun and dropped it in the pouch sewed into the case. When he zipped up the case he gave another shake of the head, which I took to be a gesture of resignation—the kind of thing a person does when he reluctantly but absolutely positively must "do his duty."

But duty to law and duty to conscience often conflict, and Officer Stickler reminded me of that by instead of giving the order for me to turn around and be put in irons, he took the zipped-up case and tossed it like a frisbee into the Bus. When it landed on the bunk in the back, he said, "That's how we carry guns through the state of Illinois," and without another word, nodded to Officer Standby, whereupon they both went to their car and drove away.

Later, when I popped a beer in celebration of their departure, I drank it in celebration of their arrival. Not for their arrival to enforce the law in the state of

Illinois, but for the arrival of their true role in the enforcement of law that conforms with the state of justice.

    End of report.

Rob Wintercorn leads a team of VW cardiologists doing emergency surgery on a #3 heart valve.

A very upscale refugee camp for those fleeing the state of the ordinary.

This here lake used to be a river until an earthquake dropped that there mountain into this here valley.

Let's look on the bright side: A bus is a lot easier to push without all that heavy gas in the tank.

If you have an oil leak at the Continental Divide, half of it will flow east to defile the Atlantic and the rest will flow west to pollute the Pacific.

There they go again - guys comparing sizes.
(Low impact mental exercise.)

Earliest attempt at bus manufacturing, mid 14th century. Notice the camouflaged warehouse on the left, although allied bombing would not start for another 500 years.

In Wisconsin, all the dogs look like Holstein cows, and we have fires in our vehicles to keep the cheese from freezing in our stomachs.

In the game of "rock, paper, scissors," why do hippies always choose "scissors"?

To ski or not to ski—that is a *question?*

Arctic shoe salesman hawking his wares.

You know you've got a good dog when he not only points
to birds but also points to fish.

There are millions of mosquitoes in Quitos, countless bees in Belize, but only one ant in Antarctica.

A hard right turn and it's all downhill from here.

Maybe it's not global warming. Maybe it's just local warming, only all over.

Angle Parking.

Solitary camping - what a bummer. Nobody to argue with when you positively know that you're absolutely right.

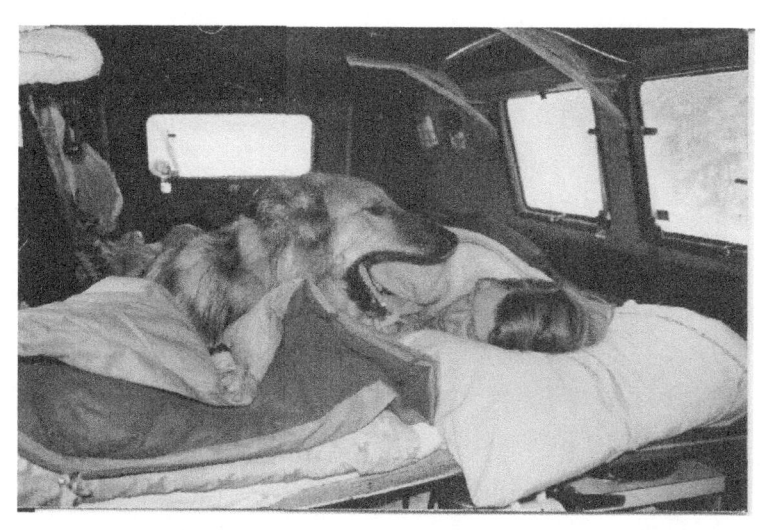

You don't have a dog just to keep the tires from drying out. He's also there for putting you to sleep with his breath and waking you up with his bark.

Whoever thinks dogs can't laugh probably doesn't
know any good cat jokes.

Deep snow and cold air—the perfect time for a blazing fire
and cup after cup of steaming hot beer.

Check the Box

"If it feels good, do it, but if it doesn't feel good, have someone else do it for minimum wage."

(Guiding principle of hippies after giving up their bell bottoms for business suits.)

Whether you're young, old, or anywhere in between, choices are all over the place. A bad choice we call "temptation" and a good one is known as "opportunity." We ask the Lord to "lead us not into temptation" and we tell ourselves, "When opportunity knocks, don't knock opportunity."

It looks good on paper, but the problem is, temptation has a better sales pitch. While opportunity usually requires an initial investment with some money down, temptation laughs at the concept of work before reward and only asks that you check the box "bill me later."

But later, when the bill comes, it comes in escalating installments. You keep owing more and more because if a dollar's worth of temptation used to fill you up, now you need twice that just to keep from feeling empty. Booze will do this, with your permission, but you know what the Big One is.

When you drive a Split-window Volkswagen Microbus, the assumption is that you are intimately familiar with the ingestion of all the chemicals that fall under the heading of "good shit." Back in the sixties and seventies, "Got any good shit? No? Want some?" was so commonly heard that it became something of a greeting. Instead of saying "Hi," "Good morning," or "How're they hang'n?" it was, "Want some good shit?"

So what does this have to do with Don? To me, Don was a reminder that some things, like God, sex, and dope, defy legislation. And to Don, I was a reminder that some things, like God, sex, and dope, are to be celebrated, and that certain signs, like the sight of a Split-window Volkswagen Microbus, are a sure indication that the driver is someone you can celebrate with.

Everyone operates according to their own reality.

This was in Georgia. In the Chattahoochee National Forest. Scotty and I were looking for a quiet place to spend the night, and wouldn't you know, right around suppertime, there it was, a quiet place, a pretty place, and a place near a little clear-water stream at the end of a winding gravel road. "Look at that," I said to Scotty, "A flat spot to park, plenty of shade, and cool running water. What more would you want?"

Well, while I was rolling out the awning and generally converting the Bus from a driver to a camper, Scotty answered that question by giving a push on my leg with his nose and backing off with a woof. In dog language that meant, "Hey, we've been driving all day. Time to change the vibe by taking a walk."

Since Scotty is one of those rare creatures who's always right without being a know-it-all, I dropped what I was doing and let him lead the way.

He started by walking up the road we came in on. But before we got to the first curve, we both stopped when we heard the sound of approaching tires on the loose gravel.

It was an old pickup, dark green and light green, and

very well kept. No rust, no dents, new tires and clean glass. It stopped when it came up even with Scotty and me, and the driver, smiling, offered his hand through the open window. "Hi, I'm Don," he said in a tone that can be best described as "Southern Gracious."

When I shook his hand and gave our names, Don looked down at Scotty and said, "Now that's a big dog."

"He's just as friendly as he is big," I said, thinking Don might need some reassurance as to Scotty's temperament.

Either at that, or for the fact that no assurance or reassurance was needed, Don casually stepped out of the truck and squatted down in front of Scotty, and as Scotty wagged his tail and licked Don's face, Don laughed and said, "Yes indeed. Three biggest lies in the world. Check's in the mail, bald is beautiful, and my dog won't bite." Then he laughed again and said, "Two out of three ain't bad."

While Don followed up his comment by using both hands to scratch behind Scotty's ears, I had a chance to scope Don out. He seemed as clean and well-kept as his truck. New running shoes, pressed shorts, and loose-fitting knit shirt, all covering a tall, fat-free frame. Age-wise, I guessed him to be late-thirties, maybe early forties.

Still, all that apparent conventionality was made unusual by his hair. His hair was pulled back in a ponytail that reached down past the middle of his back, and was tied in three different places with black leather thongs. There was something mysterious about that follicular arrangement. You could weigh all other aspects of Don's appearance, but no matter where you directed your eyes, they always came back to the sight of that ponytail, and to the feeling that the way it was worn had something to reveal.

Don was on his way to revealing what that might be when he straightened up and looked over at the Bus. "I've seen a lot of those," he said, "but not recently."

"No, not many left anymore."

"Mind if I take a closer look?"

"Actually I'd be flattered."

Don nodded and smiled and led the way over to the Bus. He kind of stalked along, as if the Bus was asleep and he wanted to get a good look at it before it woke up and drove away. Stopping in front of the open cargo doors, he let out a "hmm" at the sight of the interior, and then he turned to me and asked in a tone of true inquiry, "Do you know anything about morning glory seeds?"

"Well, I guess if you plant them you get those little flowers people call morning glories."

"You get more than that," Don said.

"Oh?"

"If you know anything about morning glory seeds, you know that they contain a substance much akin to lysergic acid." At that, Don moved in closer and said, "And you know what lysergic acid is."

"Yeah. LSD. No secret there."

"But I'm a natural man," Don said. "I eat nothing that runs or swims or has a heart. If I can find a seed from nature that verifies the effect produced by the labs of men, then I will know that what I know is true."

Not seeing a problem here, all I could do was state the obvious. "I guess if it's morning glory seeds you want—I mean stop in on a seed store. Pick 'em off the rack. They should be right next to the marigolds, if you go by the alphabet."

Don looked down and shook his head. "They're not the same. Those seeds have been altered. The government knows that the seeds of morning glories can give much more than flowers, so the government, in collusion with the seed companies, have taken out what nature put in."

"You mean like genetically?"

"And poisonously," Don said, looking up. "But some say seeds brought in from Mexico are seeds with no governmental alterations. Do you know of anyone doing this?"

"Me? No. I'm not much up on that."

"But wouldn't you want nature to corroborate what science has proposed?"

"No, I mean when I said I wasn't up on that, I meant if I did something like LSD today it would be the first time. I guess I just never got into it. Pot too, none of that."

Don did a quick look at the Bus and just as quickly brought his eyes back to me. "Oh no, no, no," he suddenly said, emphatically and apologetically. "You don't have to worry about me. I'm not from the DEA. I'm not here to set you up for a bust. You don't have to worry about me."

Even though this was an awkward moment on a controversial subject, it certainly wasn't the first time this particular theme popped up, so I just kind of shrugged and reiterated what I'd said in similar situations—that I was nothing more than a simple traveling man with a dog and a fishing pole. A tourist, you might say. Whatever went into my mouth was only standard food and drink. Typically, I liked to fry fish or grill meat and wash it down with sweet tea. Add to that coffee in the morning and a beer (maybe two) at night and there you have it.

Even after that though, and even after tacking on "Just because I drive this Bus...," Don repeated, "You don't have to worry about me," and made a case for credibility by affirming that he too was a traveling man and a simple man, although his travels were a journey not of the road but of the spirit. He believed that keeping a clean and healthy physical form was the key to having a solid base for an inquiring mind to explore the beauty and wisdom of other realms. But to visit these other realms, one needed knowledge of the "help" necessary to get there, adding somewhat conspiratorially, "I think you know what I mean."

There it was again, that implication. Split-window Volkswagen Microbus plus sixties-era driver equals intimate knowledge with improving life through chemistry. In the minds of many, that particular equation is a mathematical certainty, but to me, them thinking that way can either warn me of danger or be a form of amusement. To show Don that his assumption registered as a form of amusement, I said, "Do you know why the hippie failed his driver's test?"

When his reaction was only a cocked eyebrow, I answered with, "He kept opening the door to let out the clutch."

Even before I got all the words out though, Don was protesting. "No, no, no. I'm serious. You don't have to worry about me," he said again, and gave more evidence of his credibility by telling of the most recent time he had out-maneuvered agents sent by the Feds. He was with his six-year-old daughter and had stopped in the parking lot of a grocery store. Suddenly, three vehicles converged on him from three different directions, boxing in his truck, whereupon half a dozen federal agents leapt out with search warrants. Failing to find anything in the truck, they wanted to bring in dogs and expand the search by patting Don down and going through his clothes, but Don, dropping his tone of Southern Gracious in favor of Haughty Indignation, declined their request, stating, "I'd be pleased if you didn't, firstly because my daughter is afraid of dogs. Secondly, we came here not to be bullied but to buy groceries and nothing more, so if you gentlemen will please excuse us, that's what we're going to do." Don said it was a tense moment because he had "three big doobies" stuck down in his front pants pocket, but even at that, his ploy worked, mainly because he took his daughter by the hand as he walked away, and since a crowd was gathering, he knew that the sight of six burly bureaucrats tackling a tall and dignified father walking hand in hand with a young child would not play well in the court of public opinion.

That wasn't the end of it. Instead of gloating over the outcome, Don lamented the loss of six of his brothers. "Those agents now, they had more in common with me than their paymaster." Don went on about what a shame it was that government feels a need to control more than it feels a duty to liberate. "Those brothers of mine, would they behave the same if not for the money? If not for the gift of little pieces of green paper that have more power to blind than to buy, would their words to me be a pleasant 'Good morning' instead of a demanding 'Stand back while we search your truck!'" Don lowered his head at this, and his voice too, and when he looked up he said once again, "No, you don't have to worry about me," making the words sound all-inclusive, as if any fear, anywhere, of him, his methods or his motives would be met with a heartfelt mixture of pity and compassion for the fearful.

Here's where I offered Don a beer. It was meant to convey a feeling of two-way trust. I could take him for his word and he could take me for mine. Without missing a beat, he added to the offer by saying, "Bring that beer to my house. "We'll have it with dinner. My wife is in the process of cooking as we speak, and good person that she is, she'd like nothing better than having a like-minded guest to share it with."

I'll have to admit it was a tempting offer. But after glancing at the Bus and looking at Scotty, I hesitated. Scotty was a big dog. Don said his daughter was afraid of dogs.

Then there was the Bus. Oh yes, the Bus. If the Feds had Don on the radar, did he need something like the Bus parked at his house to add another blip to the screen?

When he sensed my indecision, he said, "I'll let you make up your mind on your own. My house is just down the road. Take a left at the blacktop and two miles later you'll see my place on the right. It's the little white house with the bright green trim."

When he left, he called back. "Doesn't necessarily have to be now. You're welcome anytime."

Instead of deciding one way or another, I stayed undecided, knowing that indecision is itself a decision, especially in matters where each alternative has a high rate of failure.

Much later, and to this day, I wondered how the

visit would have played out. If the past is any indication—and it always is—it would have played out as a continuation of what had already played. While Don would go on about his preferred way to enter the realm of the spirit, I would be thinking out loud about the human wreckage that used to be life. We'd trade stories. His would be about unity and insight and mine would be about disintegration and collapse, starting with a friend who turned into a stranger by living in a hole dug into the side of a hill and speaking in a language he recently learned while communicating with beings from another world. Still, that was mild compared to Jeedy, who left his wife and family, killed his mother, and ended up totally psychotic and finally dead.

No, one visit with Don would not be enough for me to verbally sift through that kind of rubble, and one visit with me would not be enough for Don to rhapsodize on the value of the "assistance" available when it came to exploring the reality of other realms.

In the end, it all came down to what my dad once said. We were sitting around a campfire, just the two of us. This was in a big woods, and far, far from a small world. After a long discussion on the topics of God, Sex, Dope, and Beer, and after exhausting the topics and drinking the beer, he put it all to bed by simply saying, "When anything goes, everything stops."

In Good Hands

"Whenever foresight draws a map, hindsight gives it a name."

(The last message from a failed Mars mission that ended up lost in a crack on Uranus.)

    Most of what we do gets off to a better start by first drawing up a plan. Not that forethought guarantees a good end—talk to those who plan for storms or war—but to get off to a good start, start with a plan.

    Ours was such a simple plan. Time-tested too. So time-tested, in fact, that the plan was almost more of a routine than a plan. Pack the Bus with camping gear, top off the cooler with food and drink, and fill the tank and check the oil.

    From there, it was up to the Bus. We did the planning and the packing, and now it was up to the Bus to take us where we planned to go. And where we planned to go was to a Secret Spot. That is, another Secret Spot in a long line of the same. Over the years and over the road, the Bus has camped in too many places to count, but some of those places count more than others. Much more, perhaps for the scenery or the serenity, maybe for the weather, or possibly just for the fishing. But for whatever the reason, the Secret Spots register as worthy of visiting and revisiting, and as such are kept secret, since their value, unlike a beer bottle, seems to increase when found empty.

Beano was riding shotgun. What was about to happen wasn't his fault, although what did happen was soon to take place on his side of the Bus.

We were just starting to pick up speed. We were passing through a small town where the posted limit was 25, and now, on the outskirts, we began to move back up to highway speed. Here, as usual, I was listening to the sound of the accelerating engine for any telltale signs of trouble, and here, as usual, the response from the engine was, "Don't worry, be happy."

So I stopped worrying and became happy—happy that the Bus was purring along magnificently, and happy that soon it would deliver a dad and his son to a Spot so Secret that even a bloodhound couldn't find it unless Hansel and Gretel were along and left a trail of steak and bacon.

But the happies only lasted till we got to 40 miles per hour. I remember looking at the speedo and seeing the needle reach 40, and in the same instant the right back corner of the Bus dropped to the road in the manner of a deer with its hind leg shot off.

From there it was all reflex. There was no time to question the collapse. Instinctively my foot went from the gas to the brake, but with no effect. The brakes were gone. At that, I did my best to keep the Bus on the road, no small task because the old can of sauerkraut was rocking and reeling like a boat shooting through rapids, and all I could do was flip the steering wheel this way and that to keep us from crossing the centerline or plowing into the ditch.

The ditch gave the first clue as to what this was all about. As I glanced ditchward in another attempt to keep the Bus from going there, I noticed a tire. It was rolling almost parallel with the Bus and was moving at the same speed.

It didn't take a road scholar to figure out where that tire came from. It came from the back of the Bus, and along with the wheel assembly it was bouncing and rolling along on its own momentum at close to 40 miles per hour.

That was the bad news.

The worse news was that many, many new cars were parked in the line of fire. As I fought the steering wheel to keep the careening Bus on something of a course, the free-rolling tire and wheel assembly angled out of the back side of the ditch and put its sights on a row of brand-new Toyota sedans parked in front of an automotive dealership.

Luckily there was a sign in its path. It was an old sign, a tin sign, rusted and rectangular, and held up on each end by metal poles. Almost as if responding to a command, the rapidly rolling tire did a high bounce, became airborne, and smacked directly into the middle of the sign, and the last thing I saw as we sailed on by was the sign enfolding the tire in the manner of a baseball glove catching a ball, only with a clang instead of a thud.

This outcome was made even better when the Bus finally coasted to a halt another hundred yards down the road. Nobody was hurt. For a long moment Beano and I just looked at each other with that look you give when your mind has not yet caught up with events, and then we got out of the Bus to answer questions. Other drivers had seen our struggle to stay on the road and had stopped to ask, "Now what was *that* all about!?"

One look to the rear solved the mystery. The back axle was broken. The end of the right rear axle had sheared off, along with the axle nut, leaving nothing to hold the wheel on the spline. And since the brake line is attached to the brake drum which is part of the wheel assembly and which needs to be in place for the brakes to work, this left the Bus to bounce and weave down the road on three wheels with no brakes.

But that's not what I told the gathering crowd. They deserved an answer more in tune with my feeling of relief, so I let out a big breath of air and said, "Never buy a Bus from Texas. They have that rodeo gene that always wants to buck you off."

There were nods from some and smiles from others, but mostly there was help. Together we lifted and pushed the Bus to the shoulder of the road, and once it was safely there, there were offers of more help. "Can we call you a wrecker?" someone said, while someone else went back and retrieved the errant wheel. Beano and I were still too stunned from the thought of what could have happened to worry about what to do next, so I respectfully declined any further offers of help in favor of just letting the dust settle before going ahead with any kind of decision.

Turned out, the decision was made for us, and was made by a sign on the front of a building. "Pete's Welding," the sign read. Ironically and almost unbelievably, where the Bus came to a halt was right in front of a welding shop.

So you know what that meant. Right. Lift up the back end of the Bus, push the wheel onto the spline, and roll the old can of sauerkraut up to the door of the welding shop.

When I explained to one of the guys who helped with the move that what I had in mind was to have a lump from a welding rod tacked onto the end of the axle to hold the wheel on, he, apparently a local guy, said, "Well, if Pete can't do it, then nobody can."

Others, too, had faith in Pete, since Beano and I had to wait while he attended to the needs of several customers ahead of us. When our turn came, I gave a run-down of what happened and wondered if he could take a welding rod to the end of the axle and zap it into a lump to keep the wheel in place. Yes, was his answer, but added that this method of repair should be seen as only temporary and that we should go straight home and start over with a new axle, reason being that the composition of the metal in the axle might not match the composition of the metal in the welding rod and that cracks could form where the two metals met, causing the lump to let go. I agreed with everything he said. But the problem was, I had more faith in Pete's welding than I had in his advice. So did Beano, so when we left, instead of going straight home, we went to a parking lot where we got out, crouched down by the back wheel, and closely examined the weld. The weld was very neatly done and looked like something that would still be there when time and rust claimed the rest of the Bus. The integrity of the weld and its apparent look of permanence had Beano and I wondering if perhaps maybe we could just possibly take the scenic route back home. That is, continue on to the Secret Spot, celebrate our arrival with fried fish and fresh picked wild edibles, and return at our leisure.

The key word here was "leisure." Especially when it came to driving. Even though I was dumb enough to ignore the sage advice of a welding professional, that act of stupidity was still not stupid enough. No, I needed to knock off another fifty IQ points and take to the highway driving with a Bus where "the rear axle might not match the composition of the metal in the welding rod," and furthermore, "cracks could form," and, well, you know the rest.

So we took to the back roads. We bumped along on cow paths and bike trails at a speed only fast enough to keep from getting a parking ticket. That lack of forward motion guaranteed that if "the lump let go," the loss of the rear wheel would be greeted with a yawn instead of an involuntary evacuation of the bowels.

And so it went. When the lump let go, the Bus just kind of sat down on the back corner and the wheel wobbled ahead and fell down like a drunk in a failed attempt to make it from the bar stool to the binjo.

This time when Beano and I looked at each other, instead of putting on faces of panic and confusion, we both said as one, "Now what."

And no wonder. There the Bus was, squatting on the side of a back road in the middle of farm country. Way up yonder was the sight of a silo and some other buildings, and to the back was evidence of another farm. I say evidence because from our position of about a quarter mile away, the farm looked as though it might not be occupied. The house was long overdue for a paint job, tall grass was in the yard, and the rest of the scene had that look of no recent activity.

Still, this seemed our best hope. Even if we found that the place was abandoned, at least we'd be off the road and would be in a better position to figure out what to do next.

With that in mind, we jacked up the back of the Bus, pushed the wheel onto the spline, and started the long process of closing the gap between the Bus and the farm. What made the process so long was stopping every eight or ten feet to make adjustments. Since the lump was gone and there was nothing on the end of the axle to hold the wheel on, we could only back up a few feet at a time before the wheel began to spiral off the spline. Beano would let me know when to stop. I'd start backing up very slowly, and Beano, walking backwards behind, would give a holler when the wheel was about to let go. I'd stop then, and then we'd jack up the back corner, and then we'd push the wheel into place.

After a high but uncounted number of stoppings, jackings, and pushings, we finally arrived at the farm. But once we were there, it seemed even more unoccupied than it did from a distance. Old and chipped paint was on the house, a corner of the foundation was crumbling, and when we walked up to the back door, that too had chipped paint plus appeared to be hanging a little bit off plumb.

When we knocked, it was more of a formality than anything. Neither one of us expected anyone to come to the door, but wouldn't you know, before we could shrug and walk away, a curtain parted on the other side of the glass and there looking up at us was the face of a little old lady.

We only got a glimpse of her before the curtain fell back, but as soon as it did the door opened and out stepped a large old man.

After exchanging greetings, I explained our situation and wondered if he had an electric drill, further explaining that I had a quarter-inch drill bit, and if he could be kind enough to let us use his drill, then I could drill through the brake drum, and continue on drilling through the axle until there was a hole through the both of them that I could pound a screwdriver in, which would lock the wheel in place. "You know, kind of like a dowel or something."

He listened intently as I spoke, and when I stopped he said, "I'll help you any way I can." His tone was very even and dutiful, and he backed up his words by leading the way to an outbuilding.

Walking behind, I had a chance to check him out. He was big and lean and had one of those slightly stiff, bent-at-the-hip gaits that you see on guys who are getting up there in age. I guessed him to be somewhere a little north of seventy, although the intent way he listened and the direct way he offered to help gave the impression of someone who does not measure life in years but thinks of it in terms of remaining capability. This impression was reinforced when, once inside the outbuilding, he plugged in a long extension cord and began unreeling it outside. "If you can back up your van just a little bit more," he said when coming to the end of the cord, "then I think we can reach it from here."

When Beano and I backed the Bus into place, he not only had the cord ready and the drill plugged in, but also had a bit in the chuck. "Where would you like the hole?" he said, when Beano and I met him by the back wheel.

"I was thinking right about here," I said, pointing to a flared part of the brake drum where it narrowed and surrounded the axle spline. "But I didn't mean to use your drill bit. Here, I'll put mine in, in case something happens."

"We should use this one," he said. "It's made for drilling hardened metal."

Before I could counter with "This one is also made for drilling hardened metal," he bent over and started drilling in the place I indicated.

At that point all I could do was stand there and fidget. It was one of those moments where the help you're getting seems almost painful. There I was, a dumb jerk having put himself in a tight spot, and now, instead of suffering for it, *instead of having to work my own way out of it,* someone else is doing the work, and that particular someone is in his seventies, while me, half that age, I end up fidgeting on the sidelines thinking about changing my name to Stan Daround.

It got worse—or better, depending on whether you want this show back on the road or whether you want to show that there are those willing to give you their right arm even if you already have five right arms of your own.

His wife appeared. Earlier, we only saw her face when she pulled back the curtain, but now she was standing among us, watching the progress of the drill and smoking a pipe. Her smoking that pipe almost seemed fitting when you put that next to the fact that she was less than five feet tall, round, and had close-cut stiff gray hair on a head seemingly attached to the body without benefit of a neck.

If her appearance was noteworthy, what was even more noteworthy was what she had to say. "Sid," she said. "Did you tell them about your accident?"

And then Sid, still pushing on the drill, replied, "In a minute. I'm almost done here."

"He got run over by a tractor." She motioned with her pipe to a section of the farmyard between two buildings and said very matter-of-factly, "Right over there."

Sid was standing up by then. He had the hole drilled through the brake drum and was unplugging the drill from the extension cord. "Yeah," he said, pointing with the drill to where his wife had pointed before. "It happened right over there. I was reaching for my jacket and accidentally hit the gearshift."

Reaching for his jacket and hitting the gearshift would have been nothing more than ho-hum, except he was off the tractor and standing in front of the back wheel. That made all the difference in the world, because the tractor, now in forward motion, jerked ahead, and the back wheel knocked Sid on his back and drove over him, starting at the hip and continuing at an angle across his rib cage and over his arm. "Just missed my head," Sid said.

There was more. The tractor was hooked up to a wagon. So after the tractor rolled over him, here comes the wagon, and it too ran him over. "Luckily it was empty," Sid said, "or I could have been hurt."

Thinking he was not really hurt, Sid quickly rose from his supine position, intent on chasing down the runaway tractor. But the tractor, motoring along without a driver, was zigzagging all over the place. Knowing he would be unable to approach the tractor from the side, Sid decided to tackle the problem from behind. He caught up with the back of the wagon with the idea of climbing on board and making his way forward until he could pull the tractor out of gear.

But fate found a better way. The tractor plowed into the side of a small shed and the resultant impact knocked loose the tin roof which fell in front of the tractor and ended up under the back tires, so now the tractor, still pushing against the shed, can't get any traction because the back tires are slipping on what used to be the tin roof. "So I just walked up and turned off the key."

Up till then, the tone of Sid's voice was totally devoid of drama. He could have been talking about milk and eggs on a grocery list. Here though, his tone took on gravity, and the weight of it bent his shoulders as he spoke. "I knew something was wrong. I felt kind of sick and dizzy. My chest hurt, and when I breathed I could feel things crackling inside."

His wife felt something too. "I don't know what it was," she said, "but I was doing the laundry and all of a sudden this feeling came over me. *Something happened to Sid!* When I ran outside he was sitting on the ground. He had his head down and looked really bad. What's wrong, what's wrong! I kept saying, and he said, I think you better call Marsha."

Marsha was their daughter, a nurse in a nearby town. Sid didn't remember what happened next. He didn't remember anything until he woke up later in the hospital. "Bertie was holding my hand and said everything was going to be alright. I had a cracked pelvis and a punctured lung, a slew of broken ribs and severe bruising on half of my body. But I was in good hands and everything was going to be alright."

And then Bertie, who was holding Sid's hand in the hospital, reached out and held Sid's hand again, and in that moment, nobody had to tell anybody what love was.

Bertie broke the spell. "Sid," she said almost cheerfully, "Show them your bruises," whereupon Sid, back to his matter-of-fact self, looked down and lifted the front of his shirt.

My first reaction was one of shock. There on Sid's chest was this big cleat mark. It was gray and rectangular and set at an angle across Sid's rib cage. Next to it was part of another cleat mark set at a right angle to the first. No doubt about it. These were imprints from the lugs on a tractor tire. *Big* imprints, very distinct, and looking recent.

Which caused me to ask, "When did this happen?"

Sid let his shirt down and looked at his wife. "When was that, Bertie? Was that—yeah. Yeah. Last month today. Exactly one month ago today."

Bertie was nodding and squeezing Sid's hand, but myself, all I could do was stand there and blink. Finally I said, "And you're home already?"

"They tried to keep him longer," Bertie said. "But Sid wanted to come home. He said the hospital should be there for people who really need it. Isn't that right, Sid? Isn't that what you said?"

"Yeah, but I meant the food. When you're used to the best, second best just doesn't quite cut it." At that, he returned the squeeze to Bertie's hand, and pretty soon everyone was smiling, but not because something was funny.

I let out this big breath of air and looked at Beano. I think we were both thinking the same thing. Here's this big old guy—Sid—not even half recovered from a near-fatal accident that almost took his life and left him for dead, and what is he doing? He's doing all the work necessary to get two (physically) fit and healthy human beans back on the road.

As if that wasn't enough, he doesn't stop there. Still holding the drill with one hand, and still holding Bertie's hand with the other, he very sincerely said, "Is there anything else I can help you with?"

"No," I almost said. I almost said, "No, thank you, you've already done too much," but instead I said,

"Yes, there's one more thing I'd like you to do." At that, I reached in my pocket and pulled out a twenty. "Please take this," I said.

Embarrassed, he respectfully declined, and for emphasis stepped back from the offered bill.

"Well then at least let me wind up the cord and put away the drill." I mean, I had to do *something*.

This he let me do, but before he handed over the drill, Bertie said, "We'll be going inside now. I'd like to invite you in for coffee, but I think Sid needs to rest."

Before I put the drill away, Beano and I watched them walk to the house. It was a slow walk, and a long walk, and if Bertie was holding Sid's hand before, now she was supporting it. She opened the door when they stopped at the house and let Sid go in first, putting her hand on his back as he stepped inside. And here you could see that even though Sid stood high over Bertie in height, in heart they were equal in size.

Sacred Ground

"They put clouds in the place of a sunny sky to make rain on the world of someone else."

(Common Native American lamentation often heard after 1492.)

Really, it was too soon to call it a day. The sun was still high, and there were none of those gusting crosswinds that you normally need to fight when angling a Bus northwest across the Dakota prairie. All in all, a good day to stay on the road.

But something about the sight of it made me stop. Maybe it was the contrast. It was situated in a dip in a surrounding land immensely flat. No doubt the dip had been created by falling water. Apparently this was a gathering place for raindrops, and collecting here they began to move, and as gravity coaxed them along they took some dirt with them. More dirt followed with each flow until the migration of water and dirt left a dip in the land. It was a deep dip, and broad, and at the bottom was evidence of a stream that only flowed when there was an overflow, and on each sloping side tall pines were growing, all growing in sharp contrast to the miles and miles of flat, open prairie north, south, east, and west.

Adding to the attraction was a turnout. It was near

the bottom of the dip and led to a cluster of many campsites, probably twenty in all, all spread out randomly, all under tall pines, and each having a fire pit and a level spot for a tent or a vehicle.

Wow, I thought, so much prairie and now this. It seemed like an oasis. There was no water in the stream bed, but after driving half a day in the glaring light of a sun-baked land, finding a shaded spot tucked down in a friendly hollow blessed with many campsites was like finding an oasis after traveling in the desert.

Inviting as it was though, I was the only one there. There were no other people and no evidence of anyone having been there recently. No half-burned firewood, no trash, not even a tire track.

Odd as that was, there was nothing eerie about it. Rather than looking deserted or abandoned, instead the place had an aura of patience. It was a clean and quiet patience, an inviting patience, and when I parked at one of the campsites and stepped out of the Bus this aura was added to by the delicate fragrance of cool air mixed with pine scent.

Soon the scent of pine was gone and in its place was the smell of wood smoke. I figured why not. Here was as good a place as it gets to make a stop and have a bite, so I kicked up some sticks and in a matter of minutes had a fire going and a pan of water boiling for a pot of sweet tea. Add to that a big plate of fresh meat for roasting over a pile of glowing coals and now you're talking the good life.

Making it even better was Jimmy Yellowbird. He wasn't there yet. In fact I had no idea there was such a person with such a name until here comes another VW Bus. Since the air was quiet and traffic was almost non-existent, there was the sound of the Bus before there was the sight of it, and sure enough, coming from the northwest, it suddenly appears at the top of the dip, descends, and takes a left into the turnout. While I'm shaking my head at the sheer coincidence of seeing another Split-window Volkswagen Microbus with the same paint scheme as my own, the driver pulls up next to my campfire, sticks a bronze head out the window, and with a big smile showing a set of appliance-white teeth, says, "Got room for one more Busman?"

Something about his cheery tone made me make a big sweep of my hand to all the other empty campsites. "You should've called ahead," I said. "Everything's booked up."

"Guess I'll have to stay with you then," he said, breezily, and stepping down out of the Bus and reached back to the front seat and came out holding a brown bag. "Lotsa goodies in here. Here, take this," he said. "Buffalo jerky. Don't ask me the recipe. You go by ear. One day sage, one day smoke and salt. Only the meat stays the same."

"Mmm, good," I said, after taking a bite off the end. "Thanks."

There were introductions then, a handshake and an exchange of names. "If I told you my Crow name, you wouldn't remember it. It translates into Yellowbird. Days like these though, you need two names, so I grow up called Jimmy. Oh I get *called* other names, but if you want me to answer you have to ask for Jimmy."

Jimmy was on his way to see his sister and brother-in-law. "But it's too soon to go there. They live down in Sioux country, and a guy like me, a proud descendent of scouts for Custer, it's better I sneak in at night." He winked then and added, "Easier to steal the horses."

At the mention of horses, Jimmy asked if I still had the old 40 horse motor in my Bus. When I answered that the one in there now was the second 40 after wearing out the original, he was shaking his head even before I finished. "Man, we have to bring you out of the Stone Age. Here, take my keys." He pointed out to the road and said, "Run my Bus up the hill over there and see what you're missing. Really. Go. You need to see what you're missing."

When I took his Bus up the hill and came back smiling, he said, "See? That's the new 1600. It just came out last year." According to Jimmy, if perfection was the question, then the new 1600 cubic centimeter air-cooled VW engine was the answer. It was the motor that Volkswagen should have put in its Buses from Day One. "The old 40 horse—the 1200—it was born an old nag with a bad leg. That motor gets all hot and tired climbing hills and bucking wind and has no power left to give you a good ride. You saw the difference. What do you think?

"I think we should trade Buses."

"I hear that all the time. And you know what I say? Don't make me cut my hair."

Jimmy explained that when a Crow cuts his hair, it's an expression of grief. "A long time ago," he said, "we saw that getting over real grief takes about as long as growing back a head of long hair. Hell, if I lost my Bus I wouldn't just cut my hair. I'd pull it out by the roots."

We laughed at that, both knowing that this kind of exaggeration was only another way of making it known that stuck behind the wheel in anything but a Split-window Volkswagen Microbus was like being kept in a cage and fed nothing but conformity. "Oh I've had cars and trucks," Jimmy said. "But something was missing.

Driving them was like wearing clothes made to fit someone who wasn't me."

Jimmy said he found the proper fit while hitchhiking. On the long stretch from Broadus, Montana to Spearfish, South Dakota, one of his rides was in a Split-window Volkswagen Microbus. When the driver got tired he asked Jimmy to drive. "That did it," Jimmy said. "Once I got behind the wheel I started singing." Not long after, Jimmy found the Bus he wanted and took care of it like it was his first-born son. He did all his own repair and maintenance and his wife Dee-Dee took care of the interior. "She'd be here now except she's not one for going down into Sioux country." Once again Jimmy mentioned that he was on his way to see his sister and brother-in-law. It wasn't his idea, but Robert—that was his brother-in-law—called and asked for help. Something was wrong with his van. "A Ford, wouldn't you know." And Robert, "being a Sioux," knew nothing of mechanics, so this is where Jimmy came in. "I can fix anything," Jimmy said. "Show me a problem and give me a book and I can fix the world. Oh wait a minute. I take that back. I can't fix Robert's drinking problem. Nobody can, unless he helps."

In fact, that's why Jimmy stopped at the campground. I didn't know it, Jimmy said, but this wasn't a campground in the traditional sense. This was one of the places where the tribes held their powwows, and was considered sacred ground.

At that I wondered out loud if maybe I was trespassing but Jimmy said the spirits who guard this place would not let me see it if they thought I would bring disrespect. If they thought I would leave trash or piss too close to a campfire, they would blind me "up there," Jimmy said, pointing southeast where the road entered the dip, and would let me see again "over there"—swinging his arm to point northwest where the road left the area. "So if you had no business here, on your way through you would see nothing but road."

Jimmy's business here was to prepare for the trip into Sioux country. According to him, you don't just pack a sandwich and a jug of water and drive on down. Not even a Blackfoot would do that. First you make buffalo jerky on your own ground and then you come to a gathering place to count coup. Did I know about counting coup? "The spirits of many bad men are in this bottle," Jimmy said, pulling a small glass container out of his pocket. "The glass is red, like them. Now watch what happens when I put it on the ground over there by that rock."

When Jimmy came back after placing the bottle, he reached into the cargo area of his Bus and pulled out a .22 rifle.

If I first thought he was going to load up and blast the bottle, I needed to think again. Because instead of hitting the bottle directly, each shot he fired only made the bottle jump. He sent the bullets into the dirt directly below the bottle, and the exploding dirt made the bottle flip and jump as if it were a person bumped by the business end of a cattle prod.

After several times of this, he began to explain. "Where I'm going has many hostile Injuns. They live in Sioux country, but I have their spirits in that bottle. When I shoot, those spirits are seeing great force. This is to let them know that Jimmy Yellowbird means no harm but could cause great harm."

"So why don't you just blast the bottle and be done with them?"

"When you kill your enemy you release his spirit," Jimmy said. "You don't want that. You want him here for a long time so you can both understand."

"Oh?"

Before he answered, Jimmy took two more shots and made the bottle jump both times. Then he put the gun back in the Bus and said, "Your enemy and you, you need to understand who you really are. Otherwise you think you are only you, and not made up of all other people and all other things. This takes time."

While I was taking time for that to sink in, Jimmy added, "My grandfather told me about that when I was ten. Later, when I was sixteen, I saw it also helps if your enemy likes your sister and needs you to put in a good word for him. Then later yet, when I was nineteen, I saw it helped even more when you fixed his Ford, if you know what I mean. So yeah, that lamebrain Robert and me, we get along pretty good now. But Dee-Dee says don't forget. She's like her mother. Her mother says Crow blood flows toward the heart but the blood of a Sioux always flows away."

The rivalry between Jimmy and Robert started with basketball. "Most people don't know it, but we gave up our old Indian sports a long time ago. When my grandfather was a little boy he used to listen to his father and other men talk about stealing horses and women. Back then that was our sport. We'd steal women and horses from the Sioux, and they'd do the same to us. It was all in good bad fun. Sometimes we'd just stampede the horses through the village at night to make the men mad and the women scream. I have songs from my grandfather telling of this. There weren't many rules. Basketball has too many rules. Still, it's a good way to steal women. That's how my sister got hooked up with Robert. He's this big star player down in Sioux country and when he comes up to Crow country my sister starts to dance. She watches him on the court and follows all his moves. When he sees this he asks me to put in a good word for him because I know things about Robert that he doesn't want her to know."

Jimmy stopped here and took a little time-out to look over to where he was shooting at the bottle. "If you know anything about Indians," he continued, "then you know how we feel about silence. We know how to speak without using words. So I said nothing about Robert. No good words and no bad words. It's how you avoid blame. If I say bad words about Robert, then my sister blames me today. If I say good words, then she blames me tomorrow. Here's where I take a lesson from the wind. Always be quiet unless you're only passing through."

The quiet that followed had me poking at the fire I'd started earlier. "I think this fire is just about right," I said. "If we put a pan of water on these coals we can be drinking sweet tea in about five minutes. And then there's this." I raised the lid of my cooler and pulled out a big plate of fresh meat. It was fat and juicy and cut in chunks, and ready to be shoved on sticks and roasted over coals. "This here's beef," I said, giving it my best shrug. "Not as good as buffalo, they say, but you can't have everything."

"Good thing you can't," Jimmy said, nodding. "Because trouble is part of everything, and if you have everything, then you also have trouble."

He was still thinking about Robert.

Once we got the meat going though, and once we started washing it down with lots of tea sweetened up with honey, the subject slid back to Buses, starting with the sheer coincidence of two nearly identical Split-window Volkswagen Buses converging on an isolated area of the Dakota prairie at almost exactly the same time. Well, maybe not that much of a coincidence, since this was back when three shifts at the Wolfsburg factory could not keep up with VW demand. Worldwide, Beetles and Buses dominated the automotive landscape. Already, large numbers of Buses were becoming known as "Hippy Vans," with their psychedelic paint jobs and their smoky interiors. Jimmy and I though, we still had the original paint on our Buses, known as "sealing wax red and pearl white," and as far as "smoky interiors," that wasn't our thing either. Inside, our Buses reflected the simple life of a camper and a traveling man, although I have to say that Jimmy's Bus had the added look of an interior decorated with the kind of earthy color and pleasing geometry seen in Native American art. "That's Dee-Dee," Jimmy said, gesturing to the headliner that matched the blankets on the seats and pointing to the bleached skull of an eagle resting on a circle of beadwork on the dash. "If it was just me, I'd have shop rags draped over the seats and a cutting board on the dash so I wouldn't have to stop driving to make a meal."

Driving. So much of the talk was about driving, and how a stint behind the wheel of a VW Bus was an invitation to keep going. "Yeah, you're going somewhere," Jimmy said. "But if you're going somewhere in a Bus you're already somewhere while you're going." He loved to travel, especially with Dee-Dee into Indian country, and in the space of several years they had been to almost every Indian reservation in the United States and many more in Canada. Still, there was nothing like the country of the Crow, and for every urge that took them away, two urges took them back. "Look at the sky," he said. "Other tribes might have more land, but we Crows have more sky."

With that, Jimmy's voice took on a more reflective tone. "In a little while from now the days will still be warm and the nights will be cool. Five tribes will come here then, maybe more. You'll know it's time when the grass is dry at sunset and wet in the morning. All the elders will camp down here in the pines and the rest of us will be up there in the open. The elders say, Crow, you camp with the Lakota and the two of you camp among the Pigean. And when you do, leave room for the families of the Blackfoot and the Cheyenne. That's the elders for you. They look to the next world and see peace and want us to be ready for it. So Robert, thinking he's ready for it, he camps next to my family and what do you know. He gets stuck in the morning. He gets to arguing with my sister and says he is leaving. Good luck with that, because his back tires do nothing but spin in the wet grass so there he sits like a big fat Absarokee woman. This makes him mad, but he gets madder yet when my brother Henry and I push him loose. We only tried to help but pretty soon the joke gets out that two men of the Crow Nation are stronger than 120 horses of the Sioux. That's what's under the hood of Robert's van. He's got a motor in there with the power of 120 horses, but what good is power in the front if you need it in the back? You can't ask this of Robert though, because he's just like his van, with all his power up front and nothing in the back, which if you know Robert is kind of typical him, so now that same kind of power spins the tires in his head when he hears that two Crows are stronger than all of his 120 horses." At that, Jimmy shook his head and ended it with, "I have no idea why my sister hooked up with that fool."

No doubt about it. Robert was not Jimmy's favorite person.

Still there was hope.

"Before I see Robert and before my book tells me what kind of medicine to give Robert's mighty horses, I'll first stop in and see Robert's father. His name is Douglas, but a lot of people call him Chief because he doesn't drink and would rather have a horse than a car. Douglas is a good guy to talk to before seeing Robert. So how's our Bobby, I'll say to Douglas. And Douglas he'll say, 'Oh that boy, he does not know how to wait.' Douglas always says Robert needs to learn how to wait. He says Robert was born in a time of dark days and cloudy nights and had no sun or stars to guide him so he took all of his steps looking down. You get lost looking down, Douglas says, and says if our people knew how to wait they would stay in one place and look up until the clouds were no more. Douglas says I need to know this but won't truly know this until I visit him riding a horse. 'You will trade your Bus for a horse, Jimmy,' Douglas says, 'and leave behind those stone rivers that keep taking you out of your way.'"

At this, Jimmy smiled and shook his head. "Crazy Injun. What? Give up my Bus? For a *horse?* Nah. Dee-Dee wouldn't like it either."

Getting serious again, he went on with, "Don't get me wrong. Douglas is a great guy. He'll be camping down here with the elders at the next powwow. He'll come riding across the grass on his painted horse and be here two days before everyone else driving cars on the roads. It's not speed, he says. Douglas says people only need speed when they start too late or when other people start too soon.

"Which reminds me," Jimmy said, suddenly shifting tone. "I gotta get going." He looked at the sun, now low in the sky, and said the time was just about right. "If I leave now I'll be in Sioux country right around dark. That'll be too late for the kids to see me coming and throw rocks at me, and too early for Robert to be passed out drunk in front of the TV."

We both stood up then and shook hands. I thanked him for the buffalo jerky and he thanked me for the roast meat and tea. Then he took one more look at the sky and drove away.

Years later, I was traveling through the area again, this time with someone else at the wheel of a big truck. We were "making good time," as they say, moving along at more than the speed limit. Somewhere in between feeling good about "eating up the miles" and feeling a pull from times past, I spoke of my time with Jimmy Yellowbird. "In fact, I think a little ways up this road here is where we met. If I remember right, there's a hollow up here where the Indians hold their powwows. They consider it sacred ground. We should stop there just to have a look at it."

"It better be soon," the driver said. "My back teeth are floating."

But, after several more miles, and more miles yet, and after I began to show some doubt about our whereabouts, the driver pulled over and said, "I can't wait any longer."

Soon he was back behind the wheel, and once again we were moving across the Dakota prairie, "making good time" and "eating up the miles."

Even though I was still quite certain that we were on

the right road and kept looking ahead for the telltale dip into the hollow, as the miles wore on more doubt crept in. Disorientation too, which gave way to a kind of apology. "I think we missed it," I said, blinking and looking around. "This doesn't look like where it was."

And here the driver brought it all to a close by saying, "Good thing I took that piss when I did."

Winter Will Do That

*"The way to take noise out of sound
is to convert it to light."*

(Thomas Edison's secretary when she got
sick of his profanity during failed attempts
to invent the light bulb.)

    As it turned out, they were quiet enough to sneak up on a naked woman. Ordinarily, you could hear them coming a mile away. Really. Their sound over still water could be heard more than a mile distant, just what you didn't want in a wilderness setting where the lack of sound has a special sound all its own.

    We're talking about oars. Oars by themselves don't make much noise. It's them oarlocks. Do they clunk? Do they groan? Does they squeaks? To quote my ex brother-in-law, who had a habit of answering dumb questions with even dumber questions, "Hey, is a bear catholic? Does a wild pope shit in the woods?"

    Which begs another question. How can oars give locomotion without the oarsguy going loco over the sound of the motion?

    Here you might interrupt with, "Idiot, you have a canoe. You don't row a canoe. A canoe is like a naughty child. It's something you *paddle.*"

    And to this I would answer, "So's your old man, so's your old lady," because yes, it's widely accepted that a paddle is used to move a canoe over water, and it's well known that a paddle does this quietly. But—and like the north end of a southbound sumo, this is a big but—one wimp pulling on oars can outperform two hunks pushing with paddles. It's all about the linear application of leverage and the fact that arms, legs, and back are in on the effort instead of only arms.

The more I thought about this the more I wanted to outfit my canoe with oars. There was more than one reason. First of all there was the power thing, but beyond that, rowing seemed the perfect way to troll. Think about it. You cast your lines to the back and then you stick the butts of the rods in the rod holders attached to the gunwales. Now, to make the lures move and wiggle you have to make the boat move forward. So you, sitting backwards in the middle of the boat, give a pull on the oars and at the same time watch the tips of the rods. When the rods bend back and the tips dance and stab that means it's time to let go of the oars and reel in Mr. Fishie.

So much for the theory.

Now for the practice.

Since I wanted the oars to operate without being bothered by the sounds of clunk, groan, and squeak, the oarlocks would have to be something other than the standard pin and socket. But what? Pin and socket seemed to be all that was out there. Every sport shop and every catalog had only one choice. Pin and socket. It was worse than going to the polls and seeing only the names "Clinton" and "Bush," and realizing that with this choice too there would only be a continuation of the same old noise.

Clearly, the problem of noisy oarlocks was not going to be solved by sport shops, catalogs, or by self-serving politicians. For a proper solution, a person would have to consult a higher source, a wiser source, and a source unbeholden to high-dollar donors.

In this case, the source consulted was winter. Winter is the season of snow, and one of the ways of travel over snow is with a snowmobile. And to successfully travel over this frozen white world, a snowmobile must have a steering mechanism. Going one step further, a critical part of the steering mechanism is a little thing called a ball joint. But no details will be given here as to how this little dealie contributes to the steering of the snowmobile. Why? Because the important thing to keep in mind is that the ball joint exists. It simply exists. If you think beyond the fact of its simple existence and define its purpose as a necessary part in the steering mechanism of a snowmobile, then you are not likely to think that it could have another purpose, in the same way that you are not likely to think of a Clinton or a Bush as having any other purpose than fulfilling the role of a politician, when in fact if you think of either one of them as simply existing, then you can see that a better calling for either one of them would be putting children to sleep with bedtime stories.

And so it went with the ball joint. To hold it in your hand and see it as simply existing was to take note of only its function. Its function was to simply and purely allow a specific type of motion, a ball-and-socket motion that had the stem of the socket part holding fast while letting the stem on the ball part tip or turn, or tip and turn at the same time.

It is not known how long I looked at this particular assembly. Nobody was keeping track. Nobody was keeping track, either, of the number of times I held the stem of the socket part in one hand while using the other hand to tip and turn the stem on the ball part, although I distinctly remember taking several beer and piddle breaks before finally jumping up and saying, "Yes! Of course! Why didn't I see that *right away!*"

Suddenly it was so obvious. The single ball joint had the same function as all the parts of an oarlock. Look at it this way. Use your head. I mean literally use your head. Tip your head forward and then back. See how this motion mirrors the motion of an oarlock either lifting an oar out of the water or tipping it in? Okay. Now picture yourself as a cross between a politician and Pinocchio. You've told so many lies that your little wooden nose is long enough to be carved into an oar, so all you need to do to propel a boat is look to the side and tip your head down and give it a turn. Essentially your head is part of an oarlock. Your neck is the other part. If your head holds the oar, then your neck allows the oar to tip and turn. Tip and turn is also the primary function of a ball joint, so you can see what happened next. Next I welded a clamp on the socket side of the joint. Then I welded another clamp to the ball end. The idea was to clamp the socket side to the gunwales of the canoe and clamp the ball end to an oar. Once that was done, and once a set of parallel shoulders were welded onto the socket side to have the ball end only tip forward and back and not side to side, what I had was a pair of oarlocks that really, *really,* made me smile.

It got better. Since these oarlocks were not bolted to the gunwales of the canoe but instead clamped, that meant you could trim the boat by moving them forward or back depending on where the weight was parked. You've got a big dog sleeping in the front? Clamp the oarlocks farther back. Got a fat fishing buddy sitting on the rear seat? No need to push him overboard. Just move your own seat and the oarlocks more to the front.

The seat was part of the key. Just like the oarlocks, it

was movable. It was a small, webbed creation, built for your butt to be only inches off the bottom of the boat— a good rowing platform. Foam feet held the seat in place when you sat down, but with your weight off there was no problem sliding it north, south, east, or west, and thereby giving you the instant ability to balance the boat with a simple movement of yonder buttski.

Rowing a canoe has to be the greatest maritime experience since Magellan circumcised the globe with a 90-foot Yankee Clipper. When I pulled the boat off the top of the Bus and slipped it into the water, I had no idea. With the first pull on the oars though, I had to be impressed by the huge amount of result gotten from such a small amount of effort. A canoe slides easily over water anyway, but when you enhance this easy motion with a pull on a pair of oars, what you have is an effect bordering on the giddy. "What?" you say, "This is all it takes? Shouldn't it be more like work?" But no, it's not one bit like work, even if you throw your back and legs into the process so that the canoe is tooling along at a quick trot and the speed of it all has the stern throwing up a wake topped off with spray. It's fun, is what it is. And not just ordinary fun, either, but that astonishing kind of fun that had me saying, "I knew it was going to be good, but me, conceived in sin and unfit by nature to receive the grace of God—I didn't think I deserved as much as *this*."

There was more. Besides having the capacity to funly propel the canoe at a pace equal to that of a trotting fox, the oars also gave great control. This was seen right away by being able to instantly reverse course with a push on one oar and a pull on the other.

Where control counted the most though, was in the

rough stuff. As in waves kicked up by a gusting crosswind. To paddle at an angle against this kind of stuff sucks. Gusts and waves keep throwing the bow off course, and then you, chopping away with a single paddle, use up all your fat and muscle to move it back, whereas oars—two stabilizing, steadying oars—keep you humming along in a straight line, since along with their capacity for propulsion they also act as both rudder and outrigger.

And saving the best till last, the oarlocks made not a sound. Since the only moving part was a snugly fitting well-lubricated ball tipping and turning in a tightly-built socket, the oarlock made not even a whisper. Oh, you yourself could make noise—like singing loudly in praise of silent motion, or by splashing the oars with a careless stroke—but as far as the oarlocks themselves, they made the same sound as the silent bend of your own human elbow when hoisting a mighty beer.

Little did I know that such a silence of operation would not only be something to savor, but would also be the cause of a shocking embarrassment.

Leading up to it, there I was, me, just a common road-rummy silently rowing back to the Bus. The Bus was parked under hemlocks near the shoreline of a little cove cut into the edge of a northern lake. It was a wilderness lake, and to get the Bus in close you had to drive a mile down a narrow road made of ruts and rocks, turn left onto a logging trail, and cut and thread your way through windfalls and up and down hills until finally reaching that magic spot under the hemlocks.

And there, with the Bus parked near the shore, the next thing to do is launch the canoe, and this I did, pushing off into the early morning water, calm and golden water, water like a mirror on top, and below a home for fish, fish soon caught on trolling lines, and when their number equaled the number needed for a breakfast to long remember, I caught one more and turned back to the Bus, rowing slowly, silently, and reverentially, as if breaking the silence would be on the same level as responding to the blessing "Peace on Earth" with a clueless "peace on you."

So much for lofty reverence.

Now for shocking embarrassment.

My landing spot was a little past the cove. I could see it from my position on the lake. What I couldn't see was the cove itself or anything or anyone in it, since my line of travel was close to the shoreline, and the shoreline had a dense growth of trees and brush that prevented any preview of what might be in the cove.

Inevitably the canoe glided quietly into the open space forming the mouth of the cove, and here the silence was shattered with a very loud and protracted "Ohhh!"—and then me, turning my head to the source of the sound, this is where I see a naked woman standing knee-deep in the water near the shore. Her white body stood out in sharp contrast to the backdrop of the green and brown tones of the earth, but was in view for only a moment, since she quickly covered up with a towel snatched from a branch overhanging the water.

Ordinarily, when an average member of the Guy Persuasion unwittingly enters the space occupied by a shapely and naked Person of Femality, he either raises his eyebrows in leering anticipation or dissolves into a muffled burst of adolescent giggles. Ordinarily.

But living alone in the wilderness has a way of lifting you out of the ordinary and into the empathetic, so my reaction to her shock and embarrassment was an equal measure of shock and embarrassment, and with that I lowered my head and put my hand over my eyes and muttered and stuttered something that sounded like, "Sorry, just passing through. Me go now."

Later, when I came back, she was gone. Or so I thought. But when I pulled the canoe up on shore, something caught my eye. It was a yellow notebook. Somebody crouched down under a bush on the other side of the cove was writing in a yellow notebook. The notebook was easier to see than the person doing the writing, because the writer was dressed from head to foot in dull shades of brown and green woolens that blended with the backdrop in the same way that blue blends in with sky.

Soon we were exchanging names and apologies. I apologized for the sudden appearance at the mouth of the cove and she apologized for using my campsite as a bathing area, explaining that washing up in the shallow water of the cove warmed by the sun was better on the senses than the deeper, colder water by her campsite. Her campsite was on the other side of the lake, and she was there under the tutelage of Tamarack and Lettie Song, a couple dedicated to teaching the Ojibwe value of living not off the land but with the land. Their course of instruction was intended to take her through all four seasons, each with its own challenges. Foraging was one of those challenges, and so was building shelter, but the Big One was "What will happen to my mind?" She was thinking of the upcoming winter. Summer and early fall had been a very informative and extended camping trip, but late fall was soon to come and she felt that as kind of a threat. It was like the whole world was going to sleep and she had to stay awake. It was a lonely feeling, she said, and she was afraid that several months of short days, bitter cold, and deep snow would wear away all her feelings of love, curiosity, and adventure, and leave her in a state of despair.

That certainly was something to think about, I said, but what was more likely to happen is what happened to me back when I was living through a winter alone in the Yukon Territory. That winter showed me more than short days, bitter cold, and deep snow. Much more, starting with the fact that tucked into every winter were many summers. They were mini-summers, and local summers, sometimes lasting for several hours or only for a few minutes. On a sunny day you could find them on the bright side of a hill, and if the air was still, then this was a time to take off your snowshoes and lay them on the slope in the manner of an easy chair, and from there it was only a matter of plopping down and leaning back. The sun did the rest, and this it did with such confidence that when you closed your eyes and felt the warming rays you'd think you were stretched out on a tropical beach somewhere instead of basking in a local hot spot that was surrounded by temperatures of forty below zero.

"Forty below zero?" she said.

"Well, sixty or seventy below at night, but during the day it warmed up to minus forty."

By the look on her face I could tell I said the wrong thing, so I quickly added, "But you won't get that around here. Around here you'll be lucky to get twenty below, and I say lucky because this is another opportunity to enjoy a whole bunch more of mini-summers." We talked about wood fires then, and how they had that special ability to warm not only your sweet physical self but also your spirit, and how your spirit, now warm and happy, got big on gratitude, and how that gratitude got bigger yet and led to connection, and how the connection took away feelings of loneliness and isolation because gradually or suddenly you realize you are Not Alone.

Winter will do that.

The more we talked about it the more she brightened. "I feel better now," she said. "I mean I feel connected already. I do." With that, she looked out over the lake and said, "It's beautiful. It is. Do you know what I mean?"

I nodded even before she finished, and later, when the world was quiet and the snow was deep, and when the lake had ice thick enough to hold up The First National Bank, I knew what she meant even more.

More Than Circuitry and Silicon

"Is anything made by man also preferred by nature?"

A question put to himself by Dr. Meyer Schlongbaum, architect, engineer, and entomologist, trying to make sense of the wreckage left by earth quakes, violent weather, and termites.

In answer to Dr. Schlongbaum's question, little Ginny Sandberg, age 9, considered the condition of her aging hamster and said, *"What is possible soon is inevitable later."*

To which Ginny's teenage brother Morris replied, *"So when's the pizza coming?"*

(Excerpts taken from the essay *Who Needs the Devil When Man Has Free Will.*)

Wolfgang Franke. Now there was a name for you. Especially since the "e" was pronounced, so when he introduced himself, the pronunciation registered in your head as "Wolfgang Frankie."

Here you'd think a name like "Wolfgang Frankie" would be the name of a start-up German garage band or at least the on-air name of a flamboyant American DJ, but no, this was the real name of a very serious owner of a foreign car repair shop located in Rapid City, South Dakota.

According to plan, there was to be no meeting with someone whose name registered as "Wolfgang Frankie," and there was to be no stopping at his foreign car repair shop in Rapid City, South Dakota. No, the plan was to simply roll *through* South Dakota. Right. Point the Bus westward from the Midwest, breeze through the flatlands of Minnesota and South Dakota, and putter over the mountains of Montana and Idaho. And then keep on going westward until reaching the base of Mount Hood all the way out there in the distant state of far-off Oregon.

It was going to be a long ride.

You'd think a ride of that length would be punctuated with many stops, some only to upload the coffee or download the wagpipe, but most to take in scenery and take on side-trips that make the journey part of the destination.

As it was though, this trip was to be done on a schedule. We had two weeks, so giving a casual answer of "whenever" was not a valid response when the question was, "When do you suppose you and Beano will be back?" Which is to say my teenage son Beano and I were going on another ski trip—this to fulfill a dream of skiing in the summer. Mount Hood, that big cone of snow in the Pacific Northwest had runs open until the end of July, so as soon as school let out in June, Mount Hood was to be our destination.

But "destination" is one of those words that will fool you. You think "destination," and what pops up on your thought screen is a picture of the place where you want to go. The word "destination" though, is only a derivative of the word "destiny," and as we all know unless we don't know, if destination is where you want to go, it's destiny that decides where you end up.

Starting out, all signs were good. Signs got even better about forty miles into the trip when Beano's audio player ate Beano's favorite audio tape. At the sound of the tape being chewed to pieces, a sane person couldn't help but think the audio player was having a "Happy Meal," because Beano's taste in listening pleasure seemed to have less to do with sound and more to do with smell. Some say teens have a gene for that. So when the machine ground to a halt and all efforts to free the mangled tape only ended up in breaking the machine itself, I showed outward sympathy but deep within I had nothing but praise for the phrase that says God works in mysterious ways.

Devine intervention continued as the road took us farther west and into the Mississippi Valley. This is where Beano stopped lamenting the loss of his malodorous music and took in the possibilities of the scenery outside. Here the road paralleled the river and along the way were many inviting backwaters. When we passed one that looked fertile enough to qualify as an obesity clinic for the treatment of oversized bluegills and bass, Beano brightened and said, "Wouldn't it be great if we could put a boat in down there?"

But even though we had fishing poles, we weren't packing the canoe and we didn't have the time. We had skis and we had two weeks, and two weeks only, so we continued on up the road, took a bridge across the river and started the long arcing climb up the bluffs on the Minnesota side.

I paid particular attention to the sound of the motor during the climb. The Bus was moving along at a nifty fifty in fourth gear. With the pedal to the metal, the Bus held its speed exactly. That is, climbing the long hill at full throttle, the speed of the Bus held at fifty miles-per-hour, and had no more muscle to move the needle up to 51. Good, I thought. All four cylinders are operating at maximum capacity, and now is the time to listen for complaints, because unhealthy motors, like unhealthy bodies and unhealthy minds, show their first signs of trouble when put under stress.

When the motor delivered us to the top of the bluff without the faintest complaint, our next thought was how to get past the town of Black Earth without having a flat tire. Let me explain. Black Earth is a little town located in the middle of Minnesota. The last time we rolled through there heading east, the right rear tire went flat. Not only that, but when we went to jack up the back end to change the tire, the jack broke. So we eased the Bus closer to the soft dirt at the edge of the shoulder of the road, jammed some firewood under the axle, and dug the dirt out from under the tire until the wheel was spinning free. From there it was only a matter of pulling off the flat and putting on the spare.

Except the spare was flat too.

"Now what," Beano said.

"I think we should give thanks for the three round ones."

When Beano just looked at me, I went on with, "We've got two flat tires and three round ones. What does that tell you?"

Rather than give a humble answer by passively saying, "Well, I spose we should be glad that the good still outnumber the bad,"—instead he got all active and huffy and said, "That tells me we're stuck here until those stupid tires get fixed."

Or at least one of them, and as it happened, somebody came along. A passing motorist stopped, cheerfully put the tire in the trunk and us in the back seat, and drove to a place of repair. Soon we were back to the Bus and back on the road, feeling good about that, but feeling bad about our benefactor's refusal to take any money for his help.

And now—now that we were once again coming up on the town of Black Earth, memories of the former flat put us in a mood of speculation.

Beano was the first to bring it up. "I know I shouldn't be saying this," he said, "but do you think we can get past the town of Black Earth without having a flat tire?"

I knew what he meant. He was thinking the town of Black Earth might be a jinx. To put that thought to rest, and to assure him that we would roll through the little town in the middle of Minnesota with no loss of pressure in any of the tires, I got straight to the point. "God won't allow it."

"What?" he said.

"God," I said, nodding. "It's different now, but back in the day—back before God sent down His Son with a list of the New Rules, the world was all different. Where we have one God ruling now, back then the whole works was under the control of this big bunch of piddling little godlets and godlings, and god-awful…"

"Dad…"

"Wait. Each one had their territory. At the top was the Sun god, the Moon god, the god of Uranus, Pluto, and Mickey Mouse."

"Right, Mickey Mouse."

"No, I just said that to let you know what's coming. The problem is all these little godniks had immortality. You couldn't kill 'em, so they got to thinking they could do whatever and get away with it. Say if you had a god of Agriculture, for instance, and he was pissed off at Mrs. god of Agriculture. Instead of keeping it at home, what does he do? He goes out and gets drunk with his buddy, the god of Hailstorms, and together they take it out on your crops."

"Dad, you need to…"

"Now wait. And then you, just a regular guy, you get the notion that you have to suck up to these retards. To keep them from ruining the next harvest, you end up giving them part of the crop. Back then, this was called a 'sacrifice' but today we call it extortion."

"Dad, you need to put a cork in it."

"Let me finish. So we're coming up on the town of Black Earth, right? What I'm saying is, back in the day—back before God put an end to all these so-called sacrifices—for us to get past the town of Black Earth without any trouble, first we'd have to deal with the god of Flat Tires. We'd have to offer him some kind of something to keep him from making a mess of our tires by his teaming up with the god of Crooked Nails and the god of Broken Glass."

"What about the god of Bullshit?"

"Laugh if you want, but don't forget to save some of it to laugh at yourself for failing to give gratitude to God for tossing those bums out."

Beano was shading his eyes by then, and was shaking his head. After adding to this bit of drama by giving off with a long and audible sigh, he took on the look of a prosecutor and said as if setting a trap, "Oh? And if God tossed those bums out, where are they now?"

"Washington, some of them, but mostly they're in Mexico."

Anyway, if nothing else, taking the time to slice up all that baloney and pass it off as a more expensive cut of meat did get us past the town of Black Earth with no loss of pressure from any of the tires. And from there we motored on across the rest of Minnesota and beyond.

It was near Mitchell, South Dakota where we first heard the noise. And even at that, it didn't really register as a "noise." Not at first. First it was a "sound," and to describe it you'd have to say, "What we had there was a kind of a purr. It was coming from the back of the Bus, intermittent, and sounding sort of soft-like, and might even have been soothing except for the fact that it did not belong."

This makes sense when you think of working machinery as a society of parts. In the case of a healthy Split-window Volkswagen Microbus purposely puttering down the pike, the society of parts put out a continuous broadcast of nothing but good news. If that welcome monotony is broken by anything sounding "different," then "different" is instantly synonymous with "suspicion," and with suspicion comes questions.

"Did you hear that?" Beano said, after the purring became more pronounced and less intermittent.

"What," I said, playing for time.

Beano cocked his head and said, "It's not there now but it keeps coming—there, there it is again. That."

I listened and pretended I was hearing it for the first time. Then I shrugged and said, "This Bus has parts from a Jaguar. When those parts are happy, you hear purring."

Problem was, soon we heard growling. At that, Beano and I just looked at each other. Finally he said,

"Well?"

By then it was later in the day and we were coming up on the Black Hills. "Let's find us a camping spot," I said. "We need to look at this."

Something was definitely wrong with the motor. By the time we found a good place to camp, the little power plant in the back was sounding very unmotorly. Not only was it growling, but when you let off on the gas, there was this big clank followed by a thud. It got worse every time you pushed in or let out on the clutch. Oh, the motor still ran, and still had power, but when we parked and shut it down, and when we popped the motor lid and looked inside, it quickly became apparent that what we had in our field of vision was nothing short of an automotive disaster area. The front and back crankshaft seals were leaking and each had let out a spray of oil in a circular pattern inside the engine compartment. As if that wasn't bad enough, the oil was glittering with tiny flakes of shining metal. When I shook my head at the sight of it, Beano said, "Is it serious?"

For an answer, I pulled and pushed on the pulley bolted to the back end of the crankshaft. A healthy motor only allows a few thousands of an inch of push or pull in either direction. This distance is called "tolerance." In a society of people or in a society of parts, any distance beyond the zone of tolerance registers as "trouble." The trouble that registered when I pulled and pushed on the pulley was of such a magnitude as to put it in a category comparable to that point in history when the Euro slid in value after being mired in Grease.

So yes, it was serious, since this kind of trouble is caused by a worn or burnt thrust bearing, and a thrust bearing is the last thing you remove when tearing down an engine. In other words, to replace the thrust bearing, first you have to pull the motor out of the Bus. Then you have to take that whole entire motor completely apart, starting from the outside and stripping it down, piece by piece, until you get to the middle, and there, like the last box on a truck you are unloading, is the accursed thrust bearing. So no, it don't get no more serious than that.

Even though I was 99 percent sure that the evil thrust bearing was the cause of the problem, I took on the role of a Global Warming Denier and clung to that lingering one percent chance that it could be something else. Something trivial. Like for instance the iffling pin, or maybe the muffler belt. Since there was no one around to contradict that possibility by saying, "Foo', there's no such *thing* as an iffling pin *or* a muffler belt," that meant I could truthfully answer, "I don't know," when Beano once again asked, "Is it serious?"

When a situation is serious, what you need to counter the gravity of it all is something humorous. This is called balance. In the East it's called Yin and Yang, but it's best to stick with the western term "balance," because Yin and Yang is often confused with "ying-yang," and ying-yang is a common vulgarism often used to highlight an extreme, as in "The only reason she married that ugly old geezer is because he's got money up the ying-yang."

The second-best way to balance a situation saturated with the serious is to go fishing. Since we were camped by a stream, this made the second-best way a more natural fit, and even though we came back to the Bus without fish, at least the act of fishing had us coming back with a greater sense of equilibrium.

This was added to later in the night. Right around midnight we were brought out of a deep sleep by this big spotlight. Someone was parked behind us and was shining this big glaring light through the back window.

As if I didn't know who that was. Cops. Cops—when they approach you at night—always block you in so you can't drive off, and then they flood the scene with light, and it's a light that illuminates a tactic of personal safety that says, "If I can blind you with light while I stand back in the dark, that lessens the chance of testing the efficacy of this bulletproof vest."

Not wanting to argue with that, I stepped out of the Bus, making sure that the first thing the cops saw was two empty hands and a smile.

Turned out there was only one cop, and when he spoke, his tone was almost apologetic. "I only stopped to let you know there's no camping allowed here."

Here's where I too became apologetic. "Actually we're not really camping," I said. "We're stranded, is what we are." When I explained our situation, he stepped out from behind the spotlight for a closer look, and in doing so exposed himself to look more bookish than coppish, having round glasses, a round head, and having much of what should be chest centered around the middle.

The motor lid was still up, and when he approached he unclipped a flashlight from his belt and leaned down to look inside. I was already there, and was pointing to the oily spray and the metal flakes—noting that the flakes looked even more foreboding with their glitter enhanced by the beam of the flashlight.

Straightening up, he said, "I've seen worse, but that doesn't mean this looks good. Will you need a tow truck?"

No, I said, what we really needed was time. Time and information. First of all, we were on our way to western Oregon—after starting out in eastern Wisconsin. So that meant we were about a thousand miles into the trip, not an ideal place to have a mechanical breakdown. To keep it from escalating into a nervous breakdown, we had to either find a parts store or find a shop that fixed foreign cars.

Even before I finished, the cop was writing something on a notepad he pulled from his pocket. "Here," he said, tearing off the sheet. "This guy on the top, he freelances. You can get work done for about half price from him, but this next guy guarantees everything. I'd check him out first."

All this was said sympathetically, and then he added in the same tone, "I'll have my relief stop by in the morning. If you have any trouble getting started, he can call in for help."

After we thanked him, and after he wished us luck and drove off, I put the piece of paper in my wallet and took it back out in the morning. Both addresses were in Rapid City, only a few miles away, which was definitely a plus since the motor had not used its sleepy time wisely. Instead of using the night to rest up and work out the kinks, apparently it had given up on recovery and succumbed to the lure of disability checks. Yesterday's clang was even more clangy when I turned the key, and when I pushed in and let out the clutch, the resultant thud was not only louder but you could literally feel the vibration in the seat.

Making matters worse, once we got going, that former purr that went to a growl soon had the sound of someone blowing into a kazoo while operating a rivet gun. This was the motor's way of saying, "You have something like five minutes before God takes my soul to the Big Junkyard in the Sky."

Not one to ignore a divine pronouncement, and certainly not one wanting a ticket for littering in case the motor decided to fall off in pieces as we drove along, I wasted no time finding the address recommended by the cop the night before. It was the address of an Import Service, 3280 W. Chicago Street, Rapid City, South Dakota 57702.

Somebody must have heard us coming. Because as soon as we parked in front of the shop, out comes—of all people—Saddam Hussein. Or at least someone who certainly *looked* like Saddam Hussein, although a young Hussein, like back in the good old days when Mr. Hussein and our own beloved Ronald Dumsfeld used to drink champagne together.

"Wolf" though, was the name stitched in white lettering above the pocket of his crisp blue work shirt, and when he walked up and said, "What can I help you with," he said it in a tone just as crisp as his shirt.

With a guy like that, you get right down to business. Instead of giving a folksy diagnosis by claiming, "Our motor got drunk on too much ethanol in the gas and needs some tomato juice for a hangover," I kept it formal by simply saying, "It's the motor," and followed that up by stepping out of the Bus and beckoning him to follow me to the back.

When I popped the motor lid and pointed out the evidence inside, he asked to hear it run, and when I flipped the key he quickly said, "Turn it off."

Now came the hard part. There was no escape from the fact that the motor had one bearing in the grave and the rest of itself on a banana peel. We agreed on that right away. What was still up in the air though, was what to do about it. Did I want him to tear it down to the middle and build it back up to the outside? What about a re-built? Would I want the motor replaced with a fresh one minted from brand new and gently used parts? One thing was for sure. I couldn't just buy the parts and fix it all by my sweet little self. Remember we only had two weeks, and to buy the parts, find a camping spot, and do the job in style would eat up at least half of that time and maybe even more. Having his guys fix it would take the rest of the day and most of the next. So it came down to a re-built. Pull the damaged motor, and pop in a working replacement. We'd be back on the road in three shakes. Or at least one shakedown.

Wolf had three mechanics on duty. They pushed the Bus into one of the engine bays and got to work. Beano and I waited in the office, which doubled as a sweat lodge for customers to sit and overheat while dreading the arrival of the bill.

As a welcome distraction, newspapers and magazines lay on tables next to the chairs. The most interesting reading though, was the writing on the walls. Plaques and awards were in great abundance, all in praise of a master mechanic named Wolfgang Franke and his work at Import Auto. Wolfgang himself seemed indifferent to accolades, giving the impression of one who prefers to put all his focus not on past praise but on current business, this by keeping his mechanics busy and directed, answering the phone while doing paperwork, and even maintaining a strictly business attitude when a pretty blond in a sickly sounding Fiat showed up and very flirtatiously said, "Wolf, I need you."

If we thought Wolfgang Franke was running Import Auto as anything other than a business, all those doubts were dashed when he presented us with the bill. Yes, the bill. Was it high? Let's put it this way. Did Humpty Dumpty have a great autumn? And what about Noah Zark? Did he ever have a wet dream? Anyway, high, low, or Grandma's big toe, we certainly weren't going to get back on the road by saying, "Will you take ten bucks now and the rest when I win the lottery?" No, so I reached deep down into my front pants pocket and pulled out an envelope. It was the envelope that held the money set aside to pay for our trip. Over the course of many moons the amount of money put in had reached a sum of about equal to what was needed to take us all the way out to Mount Hood and all the way back and pay for all the projected expenses in between.

    The key word here is "projected," the same kind of "projected" put forth by those arrogant pronouncements, "We will be greeted as liberators," and "The oil will pay for the war."

    Yeah.

    Lucky for us though, me and Beano didn't end up no six trillion dollars in debt, didn't kill or torture no million-plus innocent people and didn't turn all the Middle Eastern sand into Worldwide quicksand. Although we did seem to lose an arm and a leg when it became necessary to take all of our savings out of an envelope and keep stacking it up until the sum of what was in the envelope about equaled what was needed to match the numbers on the last line of our little yellow invoice.

    Beano was watching as I dealt out the dollars. Each time I plucked another bill from the envelope and laid it on the counter, he would wince with that special kind of wince you haul out whenever your future is shrinking. "There goes our gas money," said the look on his face. "And there goes our money for lift tickets." And finally, "Will there be any left for food?"

When the number of dollars finally matched the number at the bottom of the invoice, Wolfgang Franke scooped up the money and scribbled "Paid" on the little yellow sheet. And when he picked up the sheet and handed it to me, he folded his arms high across his chest and said, "You have a good motor there. It's guaranteed for 3,000 miles. I built it myself." This he followed with instructions on how to drive during the break-in period, and this too was said with his arms folded high across his chest. The way he folded those arms across that chest seemed to say more than his words. Such a stance on another person would likely register as intimidating, but on Wolfgang Franke it projected an aura of competence coupled with confidence. We were free to go.

But where? To Mt. Hood? No way in Bombay. Ironically, the cost of the new motor needed to get us there effectively cancelled the very trip itself, and here it was seen that the word "Paid" written on the little yellow invoice was essentially synonymous with the word "Screwed."

Beano sensed what I was thinking. "We're not going back, are we?"

When I didn't answer, he said, "Let's not go home yet. Let's at least go to Montana."

I nodded, but it was not a nod of approval. I was thinking. In my mind I was taking the remainder of our money and translating that into miles. Although the meager amount was easily enough to get us back home, it was hardly enough to go anywhere else. Finally I said, "If we're careful how we spend, we can go to Montana. We could go to Wyoming too, but then we won't have any money for food."

Soon we were on the road to Montana—by way of Wyoming. It made sense when you looked into our food stocks. When we counted out our potatoes and measured out our rice and then totaled that up and divided the sum into meals, that pretty much took care of the carbs part. As far as the protein part, we had two fishing poles and a .22 rifle. And since it was early June, wild veggies would be all over the place—especially asparagus. Roadside asparagus was common out west. Plus there were patches beyond the roads where you could gather enough asparagus in one hour to make your pee smell funny for a whole week.

Naturally, when you're on a bare-bones budget, and off in a galaxy far, far away, you have some concerns about getting back home. All signs were encouraging when we left Import Auto. Wolfgang's motor was operating in true German fashion. To press the accelerator was to feel an instant response. The motor had no hesitation and no misfires. And later, when we had to buck a gusting headwind, and still later when we had to climb steep mountain grades, there were no rumblings of stubborn behavior or any other indications of mechanical insubordination.

We were free to go.

Which we did, first over to Cody, Wyoming, and then up the road to Powder River pass. It was a long ride, a steep ride, and a ride that peaked at 9,666 feet. On the way up I was telling Beano about another time I took this route. "Back then it was raining below, snowing higher up, and after climbing through all that and churning a little further up the grade, the Bus did a turnaround without benefit of the steering wheel." That little trick came courtesy of the snow. "The road was a little too steep and the snow a little too deep, and wouldn't you know the Bus not only came to a stop but also started skidding backwards downhill." Here I left out the part where I shit my pants, and went on with, "The back wheel on your side saved me. It dropped off the edge of the road, grabbed, and spun the front of the Bus to where it's facing downhill. From there it was only a matter of coasting down into safer territory."— proof that God is always watching, or at least evidence that Hell is currently full, and for that reason the devil keeps turning away highly qualified applicants.

Beano and I saw snow on the way up—lots of it, and deep—but none was on the road. We parked at a turnout at the top. There were big views to the west and big views to the east. A marker was in place giving a brief history of the area while also noting that we were standing on the Continental Divide. When Beano wondered what was meant by the term "Continental Divide," I explained that if we stood back-to-back at that particular spot and he peed east and I peed west, gravity would take the contents of his bladder eastward to contaminate the Atlantic, while that same gravity would take mine westward to pollute the Pacific.

Beano though, he was more interested in skiing than peeing. To walk to the eastern edge of the turnout and look down was to see a long and steep stretch of untracked powder. When we took in the scene below, I could tell what he was thinking. Well? Don't we have skis? A run like this, isn't this what skiers dream of?

"Yeahbut," I said, in answer to his thoughts. "How we gonna get back up?"

It certainly was a legitimate question. The "run"

was too steep to climb back up even without the snow. And then there was the snow itself. Exactly how deep was that stuff? The question had me thinking out loud. "Is it deep and steep enough for a skier to trigger an avalanche?"

That little vocal musing did more to dampen Beano's desire to strap on his skis than any other combination of words in the entire Linguish Anguage. You could see it in his face.

Sensing I was on a roll, and sensing that for safety's sake I should stay on a roll, I quickly added, "We should take turns going down so in case the snow lets go then one of us can tell the searchers where to look for the body." Pausing only slightly, I put my hand on his shoulder and finished with, "You go first."

In the end, it all came down to the timing. It was just another one of those things you avoid doing stupid today so you can live to do it smart tomorrow.

Besides, we had a Bus with a new motor that was guaranteed for 3,000 miles. What was the sense of being buried under a snow-slide when we were only about 500 miles into the guarantee? No, better to make good on that guarantee by giving up all thoughts of skiing in Wyoming and motoring over to Montana for the fishing.

Beano and I first fished the Madison. The Madison is a major Montana river that originates in Yellowstone National Park. Beyond that, it's a river with a well-deserved reputation for having yin and yang by having trout up the ying-yang. Browns and Rainbows, mostly, and some Cutthroats.

A little downriver from the park, we parked the Bus on a flat spot near the water, and sooner than slowly but not as fast as quickly we had three fat Browns to go with our rice and potatoes. Still, a plate is not complete without a side order of veggies, so we boiled up a handful of asparagus found earlier.

Here's where I started talking about The Biggest Patch of Wild Asparagus in The Whole World. Beano and I were both big on asparagus, so when I said, "It's out a bit more west but really not that far," we decided to make a run for it. We laid down our dollars and counted our change and came to the conclusion that, yup, we still had enough spending green to move us another notch west and eventually back home, and we further concluded that if we saved gas by using a soft pedal on the upgrades and coasted on the way down, then we could use that savings to buy a pound of butter, since everyone knows that a plate of fresh-cooked asparagus is not quite ready to eat unless it's melting pat after pat of succulent butter.

As if it too was big on asparagus, Wolfgang's motor sprang to life at the turn of the key. By then the motor was plenty broken-in and gave the impression of an athlete trained and ready to show its stuff.

Still, we drove with a soft pedal and kept our speed at a thrifty fifty. The butter. It was all about the butter.

I hadn't made a visit to The Biggest Patch of Wild Asparagus in many moons. Still, I was quite confident that during my time of abstinence nothing much had changed. The area was remote and the patch was growing on federal land. Plus there was the added advantage of it not being reachable by road. "We'll have to walk in," I said to Beano. "There's mountains all around, but then you hike in a few miles and all at once there's this flat spot of about twenty acres." I had to pause then, and shake my head in memory of my first time on the scene. "At first you'd think it's nothing but tall grass in a big field. But then you swish on out there and I mean it's like striking gold. Asparagus is all over the place. Everywhere you go in that whole twenty acres you're surrounded by wild asparagus."

True to expectations, nothing much had changed. There was some brush and a few small trees where before there was only grass, but other than that you could not walk anywhere in that field and be out of sight of spear after spear of wild and free asparagus.

Beano and I came out of there with two grocery bags chock full. And all choice cuts. That is, if you translated our loot into beef, what was in those bags was nothing less than the choicest of choicely chosen cuts of tenderloin and boneless rib-eye. It was all about the picking. By that I mean you start from the top. You take your thumb and fingers and bend the spear at the top and gradually work your way down. At a certain point on the way down, the stem starts to put up resistance. Here's where you break it off. Unless you're a termite and have a need to eat wood, you avoid breaking off the stalk below the point of resistance because below this point asparagus stops being a tender veggie and begins to morph into the category of "building material."

That night we ate one whole bag of asparagus. We pulled out our biggest pot and filled it with a fair amount from the first bag, thinking this would be more than enough to more than satisfy our simple needs.

But wants and needs are not the same, and after that first pot was cooked and gone, we steamed up another, and another, until the bag was empty and we were full, feeling like half our body weight was asparagus and a tenth of that was butter and salt.

After spending a restless night dreaming of bloat and gout, we rolled out of bed the next morning sharing a common regret. Would that be mournfully wishing we hadn't pigged out? What a stupid question. Of course not. And since a stupid question deserves an equally stupid answer, our regret was that, at our current rate of consumption, and with our currently remaining stock, we had only enough asparagus to cook one more meal where we could truly pig out. (The key words here are "pig out," and "stupid.")

But stupid or only grossly indulgent, this was a regret that only lasted as long as it took us to hike back out to the Biggest Patch of Wild Asparagus in The Whole World. A guy has to get real. When the Horn of Plenty puts music in your ear in the form of munchies in your mouth, it's your humble duty to put on an attitude of gratitude and party till you puke.

Kidding aside, we walked back from the patch after filling only one bag. We spent enough time out there to fill ten bags and maybe even an even dozen, but most of that time was used to fill up in another sense of the word and in another way with the world, because nourishment comes in many forms, and is absorbed in many ways. Here the earth and sky act as one, and food is all around, giving muscle to memory, and value to time, so when we walked back to the Bus in the light of the setting sun, the thought of another day to come was also rising.

As if to underscore the sense of connection and continuity, Wolfgang's motor too, got into the act. Instead of the normal delayed response of only coming to life after a coaxing from several turns of the starter, the next morning the motor sprang into rotation almost as fast as I flipped the key.

Impressed, I turned it off and turned to Beano. "Listen to this," I said, giving another flip of the key.

As before—quick as a reflex—the motor sprang to life. For a moment, we only looked at each other and then I said, "I think we owe Wolfgang a beer."

In the meantime, we fished. Traveling east again, we followed the rivers. We took some time to fish the Jefferson, and took more time to fish the Yellowstone, and in both places, we celebrated the time spent there by lunching down on locally caught fish and steaming, butter-soaked asparagus picked from The Biggest Some Bidgeon Patch of Wild Asparagus In The Whole Muff Uggin World.

When the Yellowstone stopped being the Yellowstone and became the Missouri, we drove. It was time. Our two weeks were up, and it was time to go back by going forward. Night and day we stayed on the road, assisted by the westerly wind, finally turning right and then left and then right again before coming to a halt at home. And there, even before unloading our gear and cleaning up, we did a count. He put what was left of his money on the kitchen table, and I did the same, and when we counted it all out, the sum total came to a monumental 73 cents.

But the big story was not on the kitchen table. It was out in the driveway, and waiting. Namely it was Wolfgang's motor, and it was waiting to roll out the next mile. Already it had proven its ability to breeze beyond the guarantee, and already I was thinking of it as a special blend of parts—a blend that you don't see every day—"Everyday" meaning what you get when you tumble together all the available parts and assemble them in their order of appearance. Take the first ten to fall out of the tumbler and compare their ability to perform with the next ten to appear, and the performance of the first group, taken together, will likely match the collective performance of the second.

But what if you hand pick? What if you think of the parts as not mere parts, but as applicants? If twenty pistons are applying for a job where only four are needed, which motor will run best? The one where the mechanic plucked four at random, or would that be the motor where a more exacting mechanic made a selection based on which four pistons would make the best team? The answer is obvious, but to make it even more obvious, whose motor were you glad you bought?

The price of Wolfgang's motor kept going down as the miles went up. What I first thought was an overcharge gradually became a fair deal. Further down the road, the fair deal morphed into a bargain, and when the mile-meter went into six figures, the bargain became a steal.

And not just *any* six figures, either. These were six *solid* figures. Let me explain. Most air-cooled engines at the 100,000 mark are like people at the century mark. The few that are left wheeze and creak, their days of performance in the distant past. The only difference is that old age makes motors too loose, while the problem with old people is that they become too tight, as in stingy spending habits, and other forms of mental and physical constipation.

Wolfgang's motor though, maintained. Halfway into the next hundred thousand, compression was holding high and steady in all four jugs. Surprisingly, the valves needed no adjustment for clearance until way late in the game. Ordinarily you pop off the valve covers at the end of every 3000 miles and check the gap between the end of the valve stem and the foot on the rocker arm. Skip over these words if you have no interest in mechanics, but do not skip over this procedure if you are the driver of an ancient Volkswagen Microbus, because that gap, if not properly kept will lead to early engine failure, and early engine failure is often followed by divorce, depression, alcoholism, and suicide. Be warned. Anyway, surprisingly, each time I popped the valve covers at the 3000-mile mark, nothing needed to be done. I'd check, and the gap would be right on. Not until the motor clocked in at mile 91,000 did there need to be an adjustment—and then that was an adjustment on an *intake* valve—and a very minor adjustment at that. I only emphasize this because all you motorheads out there know that it's the exhaust valves that mostly need the tweaking, and they usually need it often.

Not that all these miles were trouble-free. It's just that the trouble, when it reared its fearsome head, was not the fault of the engine built by Wolfgang Franke. Going from Crested Butte, Colorado, for instance, my two kids and I were trying to outrun a snowstorm coming from the south by taking a back road north up to Glenwood Springs, and barely into the trip the motor suddenly stops. Compounding the problem, it chooses to quit on a stretch of narrow road built over a dam. Sharp curves were at each end of that particular section of dangerous road, so we all quickly stepped outside, thinking that a truck barreling around either one of those bends would not have enough reaction time to keep from smacking into the Bus.

Did I mention that this was also at night?

Night, schmight, I grabbed a flashlight, popped the motor lid and looked inside. It was a wire. A wire had worked its way off a terminal, robbing the motor of ignition. Once the wire was pushed back in place, the motor responded like a racehorse sitting on the edge of the bit and chafing at its seat.

Many miles and many months later, it quit again, and again it was no fault of a master mechanic named Wolfgang Franke. This time it was only me and Mary Teal, on a run to Newfoundland. Somewhere in the wilds of Quebec, where biting bugs are thicker than the skin on a politician and the skull on a fanatic, the fuel pump gave up the ghost. Lucky for us we had a spare, but unlucky for us, taking out the old and putting in the new entailed a huge loss of blood. Nobody got cut, but man, did we get bit. Fighting off black flies and mosquitoes made a five-minute job last four times that, and at the end I was never more happy to slam down the motor lid and get my butt back in the Bus.

It wasn't over. Since we'd left the windows open while doing the repair, brazillions of bugs were buzzing inside. Once we got rolling though, they streamed out the same windows they came in, and we didn't know whether to be awed by their sheer volume or be revolted for the same reason.

It still wasn't over. After ten minutes of travel, the inside air was nearly clear of bugs but then, gradually, here comes this smell. I did my best to ignore it, not wanting to stop and investigate until maybe winter when the problem would not involve blood-draining bugs, but would merely involve temperatures of sixty below zero.

Mary Teal though, was concerned. She tipped her head and took an audible sniff and said, "Is something burning?"

Something definitely was burning, but something about the quality of the odor had me thinking not of

Wolfgang Franke and his motor overheating, but of Bernie Case and his little tin of fried grasshoppers. That was back in fourth grade—when we were both fifteen—and when I put the smell of those grasshoppers next to the smell in the Bus, I said, "I think it's bugs. When I slammed the motor lid back there, all these poor little black flies and innocent mosquitoes got trapped inside. Try not to feel bad, but I think that smell coming up is the heat from the exhaust pipes turning the fuzzy little angels into toast."

Helping us get over our grief was the appearance of another problem. It happened two days later and had to do with sound instead of smell. Again, I did my best to ignore it, since we were in Nova Scotia by then and all I wanted to think about was food. Seafood, namely, because Nova Scotia is a Maritime Province and everywhere you go are these big friendly signs in front of eateries inviting you to stop in and stick a fork into freshly steamed mussels, fried fish, and boiled crabs.

As before, it was Mary Teal who brought the problem out in the open. We were in Newfoundland by then, and with the noise getting louder she wondered if something was "wrong."

"I know what you mean," I said. "Now listen to this." We were heading into a curve, and as I turned left to stay on the road, the sound went from a burr to a growl. "Compare that with this," I said, when we headed into an opposite curve. Sure enough, instead of the burr going into a growl, when I cranked the wheel to the right, the sound disappeared altogether.

"So?" she said.

"So that means we got us a worn wheel bearing. And that also means it's on the right side of the Bus. Notice when I add weight by turning left the bearing hollers, and when I take the weight off by turning right it shuts up."

I changed out the bearing at a campground in St. Johns. By rights Mary Teal should have done the job, since the bearing was on her side of the Bus, but the thing of it was, on that particular day we were faced with two chores: laundry, and remove and replace the bearing. Of course, anyone with even one lick of sense can see that these jobs are, shall we say, "gender specific," and as such we were able to silently settle the issue of who does what without Mary Teal calling me a "male chauvinist pig," and without me countering by calling her a "female chauvinist sow."

(Just another one of those deals where instead of PC meaning "Politically Correct" it really stands for "Practical Consideration.")

Subtleties aside, Wolfgang's motor kept right on truck'n. Not even a near drowning would stop it. That happened the following winter. It was a cold winter, a long winter, and just the kind of winter you need to drive out on lake ice with no worry of breaking through. Breaking through sucks. Even partially breaking through is a total bummer. Just ask My Very Own Best Dog Freck. He was with me when we tried to make an end run around a snow-bank that was blocking an ice road on one of our favorite fishing lakes. From past experience, I knew the end run was taking us over ice rumored to be thin because of an underlying spring, and wouldn't you know the rumor turned to reality when the right side of the Bus suddenly crashed through the ice and had us tilted at an angle that made me reflexively think, "One half of one more degree and this Bus will be on its side at the bottom of the lake."

Very carefully I took myself out of the Bus. This

was no small feat because the Bus being tilted the way it was, me opening the driver's side door was like lifting a hatch, and not a light one either.

Once I got out though, and once I opened the back hatch and pulled out Freck, it was time to stop thinking about our own rescue and time to start thinking about saving the Bus.

Maybe some other time I'll go into detail about how that gig played out, but suffice it to say that, once again, this bit of trouble was no fault of the engine built by Wolfgang Franke. No, if you follow the facts and examine the evidence as to why the Bus dropped into the lake, it's easy to see who was to blame. In fact, the identity of the guilty party is so glaringly apparent that it's a total waste of words to even elaborate. Once you've rightfully eliminated Wolfgang Franke, and once you realize that Freck was certainly not at fault, really, truthfully, who's left to blame? The government, of course. Throw the bums out.

More trouble came up year after year and place after place, but time and time again the motor built by Wolfgang Franke stayed above the fray. His work seemed faultless, almost permanent, and whenever the fit was right I'd bring up his name and pour on the praise.

As much as I wanted the praise to stay praise, still, occasionally, the praise had a way of devolving into grist for gossip. Especially if you were talking to Bill, and he was with Tom. Tom and Bill are Bus guys from way back. Genetically they were German—square heads, beer-barrel bodies—but culturally they claimed to be nothing more than two country boys who loved oldies rock, wrenching on air-cooled engines, and drinking cold beer on warm summer nights. My kind of guys. Tom and Bill made a good team if only for the fact that Tom was hard of hearing and Bill took the time and effort to fill Tom in on any part of a conversation that he might be missing.

As if that wasn't quite enough, along with keeping Tom up to speed, Bill had this diabolical habit of sabotage. For him, the ultimate in humor was to relay to Tom a misinterpretation of what was said. It was all meant to make the one who said it squirm like a finalist in a urine-retention contest. This showed up big time when the talk got to engines and little old me came across as too gushy in my praise for the precision and ability of Wolfgang Franke because when I finally wound down and Tom asked Bill what I said, Bill aimed a thumb at me and deadpanned to Tom, "He's queer and his boyfriend's name is Wolfman Frankie."

Even though there were denials on one front, there were no denials on the other. Wolfgang's motor just kept on going. Twenty below zero in the sun or 106 in the shade, one flip of the key and you were on your way, rolling over most of the states in the Union, all the states in the Confederacy, and every province in Canada. Up to the Far North to fish for pike as big as pigs or down to the Deep South to search the swamps for the fabled Ivory-peckered Woodbiller, no matter— one flip of the key and you were on your way.

Beano and I got to talking about that. That was at the kitchen table, not too long ago. Somewhere into our second or third beer, up comes the time we made a run for Mount Hood and ended up broken down in South Dakota. All the details were still fresh even though Beano was a teen at the time and now his own kids were soon to be teens. We laughed, and it was that kind of laugh you laugh when you set yourself up as the butt of your own joke. Well, why not, because as it turned out, not once did we lament the fact that we failed to reach the high and mighty Mount Hood. Our talk was all about other heights, heights you get from successfully dealing with second best, and having so-called second best take you to unexpected elevations, elevations that continued for more than twenty years, partly as a result of the precision and ability of You Know Who.

"I think I owe him a dinner," I said to Beano.

"You keep saying that."

"Really. The next time I swing through Rapid City, I'm going to stop and look him up." Then I added, "I wonder if he's still in the same place."

"Well," Beano said, somewhat theatrically, "I think I can tell you right now." At that, he put down his beer bottle and reached into his pocket and pulled out this rectangular little piece of glass and plastic. It was the size and shape of a wallet and contained not money but information. I keep forgetting about that. I keep forgetting that I'm a rusty old relic from the Iron Age while Beano was born in the time of Bright and Shining Silicon. Beeg difference, Mon. Beeg.

Anyway, after looking down at the thing with his eyes and walking around on the thing with his fingers, Beano brought his eyes back up and said way too matter-of-factly, "He's dead."

"What?" I said.

"Two years ago."

For a long time I just sat there blinking, and then said as if thinking out loud, "He wasn't that old."

"Sixty-one," Beano said, looking down again.

"Sixty-one isn't that old. I wonder what happened."

Beano's little device had no ready answer for that. No doubt a little coaxing and commanding could call it up. It was good at "how, what, and when." All that data stuff. Beyond that though—when you got to the "exactly *why* is the how, what, and when," you needed more than circuitry and silicon. You needed history and motive and conscience. In Wolfgang's case, what was his history, and what kind of motives and urges of conscience had him building things that kept right on going even after he stopped? Then too, it wasn't just Wolfgang. You see it all around. The impact of motive is everywhere, and lasting, and so is the effect of conscience. When you see the benefits of this, and the after-effects and the blow-back when it's misdirected, you can't just sit there. You can't just reside in someone's pocket like money and data. It's better you ask yourself and all the world: when your words and deeds outlast your own life, what do you think they should be?

The Dual Nature of the World Outside

"Never punch someone whose face is tougher than your fist."

(Warning from Stalin to Hitler on the eve of the German invasion.)

It ain't what you don't know that gets you. It's what you don't do with what you know. They say this is doubly true of first-hand knowledge—that personal kind of hard-wired head-sense that comes stamped with: Do Not Forget.

So we should have known.

Actually, we did know. And we got that knowledge early on. At the age of four, Judelaine crashed through river ice and suffered an instant whole-body immersion. And there at the bottom of the river she was destined to be no more than an aching memory, if not for Steve. Steve was an older brother whose purpose in life seemed to be teasing his little sister until he felt a higher calling.

"I looked up," Judelaine said, "and saw Steve's hand reaching down." It was a strong hand, a determined hand, and when that hand pulled her out of the frigid water and into the frosty air, she was quickly taken home, warmed instead of mourned, and from there was able to continue on with all things that little kids do, including breaking through the ice again a few years later, although this time she was bigger and stronger and was able to come to her own rescue.

I compared that with the first time I went in. I was probably nine, exploring thin ice alone, partly out of a love of the free and solitary life, but mostly for the fact that some of the more enlightened parents in the neighborhood kept me away from their kids, based on the rumor that I tested positive for making armpit noises at the dinner table, and based on the fear that my early love of beer might spill over and corrupt their child's early love of chocolate milk. Yes, that and the fact that I wasn't baptized. So I was damned in both worlds.

Thin ice though, is not so discriminating. Its only consideration is weight. As long as you can meet the weight requirement, thin ice will always open the door to the deep. I don't know how deep the water was where I went in because before my feet could find the bottom, reflex had my arms outstretched and holding me right about the level where the belly stops being the belly and starts being the ribs. From there I scrambled back out and made a beeline for home, not an easy run because the inside of one boot, loosely tied, was full of water and weighted, while the inside of the other boot, tightly tied, was light and dry. So that made me feel like some sort of human muskrat, humping along while dragging a trap.

The big feeling though, was fear of parental misunderstanding. I was absolutely certain that when I got home, dear old mum and dad would be unimpressed that I was dry from the ribs on up, and would direct all their drama to the wet and frozen clothes below, *totally unmindful* of The Miracle of the One Dry Foot.

So very quietly and very sneakily, I snuck my wet and frozen little self down into the basement. The basement had an old furnace that radiated more heat down there than it pushed upstairs, so that's where I stood, doing a slow rotation in front of the furnace until my icy clothes were thawed to the point where I could strip them off and flap them dry.

Drying off was a long process, and by necessity a quiet process, and such a protracted endeavor gives ample time to think, and during it all I thought back to the initial shock of the breaking ice, the panicked scrambling and the desperate run home, and after that scene had been mentally replayed too many times to count, it left me with a lesson that most nine-year-olds already knew by the tender age of eight: never tell your parents anything.

Oh, I told them later—way later—at that age when parents no longer represent the Statue of Limitations. I told them of the three other times of falling through too, two dumb times and the third excusable in that I broke through while pulling out a drowning dog.

So, with all that high-impact polar plunging in our own personal histories, you'd think both Judelaine and I, battle-tested survivors in the War Against Icy Winter Immersions, would take every precaution to keep from reentering the fray.

You'd think.

But caution does not always exist alone. Sometimes it has to compete with enthusiasm.

Enthusiasm was in full flower that early spring day, mostly because we were in the process of moving up to the cabin. The cabin was built near the shore of a pretty lake tucked into the heart of the North Woods. Like most dwellings, it started out as a hole in the ground, which soon acquired a slab and then concrete walls.

Once a floor was over that, big trucks came in with big logs, and a crew that knew how to combine art and utility took each log and notched and carved and planed and cut until the whole bunch of them built one on top of the other were no longer single pieces. Now they were a log home, a place designed to be both cozy and roomy, and a place oozing with invitation, so you can see why Judelaine and I were living in a state of Happy. Soon, life in the city would be a thing of the past. And like the past, the city would be an occasional place to visit but we wouldn't want to live there.

Since the cabin still needed much exterior work and more than just a little interior tweaking, the Bus was our temporary living quarters. We set it up as restaurant, lounge, and motel and parked it under the pines overlooking the lake.

Surprisingly, the lake still had ice even though the sounds of birdsongs were everywhere and buds were beginning to open in the trees. To step out onto its stubborn surface at that time of the year made you think of Confederate money. In a perverted sense, it was as if Old Man Winter had all his money in the form of ice and only had to hold out through summer and fall before the North would rise again.

Silly as that sounds, it was no joke that even though winter was losing the war, in its retreat it would still be able to deliver a painful parting shot.

A lightly traveled country road circled the lake. To the east it was close, and you could see the lake from the road. Otherwise it was distant, and a person passing through would have no clue that a deep bowl of clean water was just beyond view, since trees ringed the lake, tall pines, oaks, and maples. Birches were there too, and poplars, and the sight of them in any season was a reason to take a leisurely walk around the lake and be impressed with the impression that nature makes the best of art, although the blind will argue for the old masters or for the more recent splattermasters and their unrivaled talent in depiction of a used food fight.

Ordinarily a walk on the road around the lake took a little less than an hour or a little more than that, depending on whether or not it was the season to stop and pick wild edibles.

However, it being early spring and with ice still on the lake, Judelaine and I thought why not take our dogs and walk the perimeter of the lake on the ice instead of taking the road. Yes, why not. The ice would soon be gone and this might be our last chance to walk the edge of the lake before the only way to travel there would be by boat. The emphasis here is on "last chance."

Right away there was a warning. Slush. Where we stepped out onto the ice was an ominous layer of thick slush. When my boot sunk several inches before hitting ice underneath, this made me think out loud, "If the water in this stuff is leaking down through cracks, well…"

Twenty steps later we were walking on solid ice.

That's because we turned south. South shore ice is always the strongest. Since this is the shore that gets the least amount of sun, ice forms here first and leaves last. So, there are times when you can safely walk the south shore while the ice on the north shore has not yet formed or is as rotten as week-old road-kill in July. To the unwary, these conditions are like bait in a trap.

What made us unwary was the persistence of satisfaction. We had finished another day at work on the cabin and were using our little walk to unwind and reflect. And now, with firm ice underfoot, any sense of foreboding was replaced with the glow of getting things done. That and a sense of mystery and gratitude. Weren't *we* the lucky ones. How did we ever *find* such a place? And after all those years alone, how did we find each *other*? Even the dogs got in on it. While we laughed, planned, and reminisced, Froggie and Scotty bounced and frolicked nearby. Those two, too, were always happy to kick along at the close of the day—and playfully, like a couple of kids finally let out of school on a Friday afternoon.

 The lake was in the form of a long oval. Cranberry bogs lined the edges. Ice was in the bogs on the south shore, but as our walk took us around the far curve and we were now walking north, the ice in the bogs gradually gave way to open water. Rather than taking this as a warning to tiptoe backwards onto safer ice, we decided to compromise by stepping farther out on the lake, knowing that melting ice is always stronger away from shore.

 Judelaine went down first. First there was this ominous cracking underfoot, and before we could step one way or the other, she was plunging downward.

 Since we were side by side, I soon followed her. She was something like a half second and a two foot drop ahead of me, but we both managed to stop at about mutual armpit level, partly for the fact of the initial buoyancy of winter clothes and partly out of extended arms grabbing floating chunks of ice.

 There was only one thought. Get out. There was no feeling of cold. The shock of the sudden immersion cancelled all feeling of cold and focused all thought on getting out.

 A quick look around revealed the dogs to still be on

top of the ice. This would change as Judelaine and I used our fists and elbows to smash our way to shore. We were about twenty feet from the edge of the cranberry bog and needed to break our way there. The ice became thinner and weaker as we smashed and swam our way in, and the dogs, looking somewhat confused but ever faithful, stayed close—too close—and suddenly they too broke through.

By then we were at the edge of the cranberry bog and could pull our soaking selves partly up. We were still standing waist deep in frigid water, but at least with some footing underneath we could reach back and pull out the dogs.

It wasn't over. We hauled Froggie out first, but something about the desperate way we pulled out the very heavy and sodden Scotty made Froggie freak, causing him to take a big leap back out onto the ice. Do not wonder why he is called Froggie.

By then the cold was really starting to penetrate. Realizing that we were absolutely, positively in a rapidly deteriorating situation where there was no time for inefficiency or do-overs, I very sweetly called Froggie to the edge of the ice, grabbed him when he crashed through, and dragged his ass back into the cranberry bog. After handing him off to Judelaine, I floated Scotty through a pool of open water in the bog, and from there the four of us waded on up to higher ground.

Of all the places to fall through the ice, we picked the spot farthest from camp. So that meant travel time back was about equal whether we took the north road or the road to the south.

We took the road to the south, knowing that this route could be shortened by leaving the road at the halfway point and hooking up with a path cutting through the woods. Every little break helped. At this point my hands had gone numb and my legs were starting to wobble.

It was dark when we found the path. Less than half a mile to go. I was starting to shake. Judelaine and the dogs were holding up magnificently, but I could feel my own time running out.

Scotty was the next to feel the effects. He and I were about the same age. If you added up his dog years, that number coincided with my own measure of time, so on the final part of the march—the last leg where the path curved steeply upward—he collapsed. I knew he wouldn't be able to get up unaided any more than I would, so I planted my feet and slid him into a position where I could lift him without falling down myself. From there, Judelaine pulled on his collar and I pushed on his butt until we got him to the top of the rise and he could walk on his own.

The Bus was straight ahead. The struggle was over. In the morning, we woke up in a warm and dry place to see the world outside covered with a blanket of newly fallen snow. It was deep and clean and clinging to everything, including our minds. That's the thing about winter. Winter can stop or move you. So much threat and so much beauty, and as we took in the dual nature of the world outside, we didn't know whether to be struck by its capacity to kill or be impressed with a beauty that makes a person glad to be alive.

Acknowledgments

A special thanks to the light that lets in the warmth, and an added thanks to the shutters that keep out the cold.

Family too, needs to be noted, especially the ones of Here who help Now, so the ones of There can help Later.

And let's remember the dogs, and how they show that loyalty is at its best when connected to conscience.

Also thanks to prison, the one with iron bars and the one of iron will, and how the thought of either can keep us out of the other.

Saving the best till last, many thousand five hundred thanks to Judelaine, who is both the light on the pathway and the beacon on the hill.

Made in the USA
Coppell, TX
15 December 2024